# THE DISCOVERY OF GOD

# THE DISCOVERY OF GOD

HENRI DE LUBAC

*Translated by*
ALEXANDER DRU

HENRY REGNERY COMPANY
CHICAGO

ENGLISH TRANSLATION © DARTON LONGMAN & TODD, LTD.
& P. J. KENNEDY AND SONS, 1960

Regnery Logos Edition 1967
By arrangement with P. J. Kenedy and Sons


First published in France by Aubier, Paris, under the title
*Sur les chemins de Dieu*


NIHIL OBSTAT: Adrianus Van Vliet, S.T.D.
                Censor Deputatus

IMPRIMATUR:   E. Morrogh Bernard,
                Vicar General

Westmonasterii, Die: 20a Januarii, 1959

# CONTENTS

## NOTE

This is a translation of *Sur les chemins de Dieu* published by Aubier (Paris, 1956), itself the third edition, recast and greatly expanded, of *De la Connaissance de Dieu*, which appeared in 1945 and 1948 (Editions du témoignage chrétien). The reader will find in the Postscript some explanation of the spirit in which the new edition has been prepared. With the author's permission, the notes have been substantially reduced. In the French edition many of the texts referred to are quoted in full. The reader who wishes to pursue any particular point might refer to the original so as to avoid the possibility of misunderstanding.

*Note*

As for me, I feel that by far my most pressing duty to God is to speak of him in all that I think and all that I say.

St. Hilary, De Trinitate, I, xxxvii.

I have never written anything on the real purpose of my endeavour.

Plato, Letter vii.

What men lack most is knowledge of God.

Fénelon, Sentiments et avis chrétiens, I.

Forte substomacharis, si adhuc pergimus quaerere: quid est Deus? tum quia toties jam quaesitum est, tum quia diffidis inventum iri. Dico tibi, pater Eugeni, solus est Deus, qui frustra numquam quaeri potest nec cum inveniri non potest.

St. Bernard, De consideratione, V, xi, 24.

# OUR KNOWLEDGE OF GOD

As chapel ended, one of the boys turned into the recreation ground and began ridiculing the sermon he had just been subjected to. Like many another, it had been a poor one. In his attempt to say something about God, the preacher had fed his youthful congregation on a flow of abstract formulae and pious platitudes which produced the most absurd effect on those whose minds were not entirely dulled. The master in charge, who was a man of God, called over the young scoffer, but instead of scolding him simply asked him: 'Hasn't it ever occurred to you that it is the most difficult subject to speak about that you can think of?' The boy was no fool. He pondered the question, and that incident became his first conscious realization of the mystery, his first contact with the twofold mystery, of God and man.

Although the thoughts in this book differ considerably from many sermons, they will not necessarily seem less ridiculous. They are deliberately fragmentary, and make no claim to replace the classic treatises on the subject or even to supplement them. They are marginal notes which make no attempt to gloss over the salutary sense of embarrassment which overcomes the mind on such an occasion. No less deliberately, they refrain from crossing the threshold of the mystery where the spiritual life is nourished, the life hidden from the eyes of the world, the intimate history of the Church which is beyond the reach of the profane. Their inadequacy, their very awkwardness may, however, provoke the reader to thought. May his reflections lead him beyond the realm of words and of human thought to find God!

# ABYSSUS ABYSSUM INVOCAT

WAS Moses right, or Xenophanes? Did God make man in his image, or is it not rather man who has made God in his?

Appearances, certainly, are on the side of Xenophanes—yet it is Moses who is right. And at bottom Xenophanes agrees. For they are not speaking of the same God, or of the same image; which is why the argument seems unending. In fact, Xenophanes has no intention of denying the divinity; on the contrary, his purpose is to recall man to the divine when he loses himself among the gods he has fashioned. In this the Christian can only approve[1] the 'intellectual revolutionary', and reckon him among those who have 'blazed a trail' to the truth.[2] His contempt for anthropomorphic gods conceals a very important positive lesson, and in effect his words evoke man's secret sympathy. They release an energy not at first distinctly understood, but which inevitably leads him on, far beyond the denial of his gods. But let us take another example, nearer our own time, in which his meaning is echoed:

> Monde, tout le mal vient de la forme des dieux. . . .
> Pourquoi mettre, au-dessus de l'Être, des Fantômes?[3]
>
> (World, all the ill comes from the form the gods take,
> Why put Phantoms above Being itself?)

Away, then, with all the projections, sublimations and

1. Cf. Clement of Alexandria, *Strom.* vii, 4 (PG, ix, 427).
2. Werner Jaeger, *The Theology of the early Greek Philosophers*, 3rd ed. (1952), p. 49; cf. p. 41.
3. Victor Hugo, *La Légende des siècles*: Le Satyre.

creations of our passions or our dreams, of our fears or anger,
of our nightmares or desires! Away with the gods who 'seem
to have been invented of set purpose by the enemy of mankind, in order to sanction crime and turn the divine to ridi-
cule'![4] Away with the gods of nothingness which leave us to
ourselves and keep us in bondage! Away with all false gods!—
for God is indeed the God of whom Moses speaks, a God
without countenance, a God who is the negation of human
gods. For the God who denies the gods of our desires is none
the less the sole God of human desire. The God before whom
everything is as though it were not is none the less the sole
God of all that is noblest in man.

All our representations of the divine are woven of elements
taken from our world, whether from the natural or the social
world; but there is a faculty or power in man which always
drives him on beyond them: the power of reason itself. That
is because in his most intimate being, made in the image of
God, there is always something which he is quite unable to
represent to himself, though he is not without experience
of it. 'He bears within himself a source of wonder, a source
of infinite self-transcendence.'[5] And that, in the last analysis,
is what allows him to know God in truth. *Abyssus abyssum
invocat* (Abyss calls to abyss).

God, Moses says, made man in his image. To which
Christian tradition adds that man, made in the image of the
incomprehensible God, is ultimately incomprehensible to himself. 'Who can enter into himself and understand himself?'[6]

---

4. Fénelon.
5. Fénelon, *Traité de l'existence de Dieu*, Part I, cap. ii, No. 52.
6. Among a host of others, St. Augustine, *De Symbolo* c. i, No. 2 (PL, xl,
628); or again Bossuet, *Élévations sur les Mystères*, 2e Semaine, 6e élévation,
on the Trinity: 'God has mingled with our souls, which represent him,
something incomprehensible. . . . I am an impenetrable mystery to myself.'

'Our spirit bears the imprint of inscrutable Nature through the mystery within it.'[7]

> *Der Abgrund meines Geistes ruft immer mit Geschrei*
> *Den Abgrund Gottes: Sag welcher tiefer sei?*[8]
>
> (The abyss of my spirit calls for ever with a cry
> To the abyss of God: Tell me, which is deeper?)

And so one cannot say that this knowledge, at its root, is a human acquisition. It is an 'image', an 'imprint', a 'seal'. It is the mark of God upon us. We do not construct it; we do not borrow it from elsewhere; it is in us, for all our misery; it is our very selves—more, even, than ourselves. It comes before the operation of will and intellect, presupposed by consciousness itself, and our initiative goes for nothing. And so it is true, indeed indispensable, to say: *Auctor nobis de Deo, Deus est; non nisi se auctore cognoscitur*[9] . . . *Deum scire nemo potest, nisi Deo docente*[10] (God himself is our authority about God; otherwise he is not known. . . . No one can have any knowledge of God unless God teaches him).

That does not mean the suppression of our natural activity of mind; it indicates the prime condition and guarantee of its validity. It does not mean substituting another principle in the place of reason; rather it means digging down to its foundation. Going back, and up, to the source. It means saying, with St. Thomas, and according to the teaching of St. Paul, that God, the creative God, manifests himself to us through his works as in a book, and that he is, moreover, the

7. St. Ephraem, cf. Edmund Beck, *Die Theologie des hl. Ephraem* in *Studia anselmiana*, 21, pp. 98, 52. St. Gregory of Nyssa, *Of the Creation of Man*, xi (PG, xliv, 156B).

8. Angelus Silesius, *The Cherubic Pilgrim*, I, 68.

9. St. Hilary, *De Trinitate*, V, i, 21 (PL, x, 143).

10. St. Irenaeus, *Adversus Haereses*, IV, vi, 4 (PG, vii, 988).

principle of the knowledge which we have to acquire by the
exercise of natural reason:

> He has put an eye into their hearts
> To show them the greatness of his works.[11]

The initiative is a double one, and our most natural and
most spontaneous activity is no more than a response. If the
reason which enlightens us were, from the first, enlightened
as to its own nature, it would be driven to make the mystic's
cry its own: 'Behold me, O my aim and end and meaning!
I cry to thee . . . Yet, no, it is thou who callest me!'[12]

Whatever the order in which things are set, God comes
before everything. He goes before us on the road, and is
always there before us. On whatever plane, it is he who makes
himself known to us. It is he who reveals himself to us.[13] The

11. *Ecclus.* 17:7, cf. Stanislas Lyonnet, S. J., *Quaestiones in Epist. ad Rom. I*,
(1955).

12. Hallaj, *Qasida I* (Dîvân, tr. Massignon, 1955, p. 4).

13. I have been asked by one or two friends why I use the terms 'reveal',
'revelation' in speaking of the natural knowledge of God, and not the
knowledge which comes to us from positive, supernatural revelation.
There are three reasons:

(1) Because the terms express an important traditional idea. Cf. St.
Irenaeus, *Adversus Haereses* II, vi, 1 (PG, vii, 724c); St. Maximus the Confessor, *Ambigua*: the visible world, 'that unique masterpiece, in which God
makes himself known by a silent revelation' (PG, xci, 1328A), etc. The
term has been transmitted even to certain manuals. It is more than authorized by St. Paul, *Rom.* 1:19: the ἐφανέρωσεν, which corresponds to the
Vulgate *manifestavit*, is often translated in older versions by *revelavit*. For
example, in the Middle Ages, William of St. Thierry: *Letter to the Brethren
of Mont-Dieu*, II, 21 (PL. clxxxiv, 352A), and Peter Lombard, *Sentences*, I,
dist. 3. Or in St. Thomas *In Epist. ad Rom.* c. i, lectio 6. Cf. J. A. Möhler,
*Symbolik*: 'We derive our knowledge of God from two sources: natural
revelation and supernatural revelation.'

(2) Because the word seems to be the best, or rather the only correlative
to the word 'image', itself biblical and traditional. Cf. among many others
St. Bonaventure, *In 2 Sent.*, dist. 16, q. 1, a. 2.

working of reason which carries us to him—not to him so
much as to the threshold of his mystery—is never but the
second wave of the rhythm which he himself has set in motion.
Whatever explanation may be given of knowledge—and St.
Thomas's explanation is not in every respect that of St.
Augustine or St. Bonaventure, for example—traditional
philosophy is unanimous on that point. In the intimacy of the
spirit, God is always the 'illuminating light' of our 'illuminated light'.[14] He is 'that uncreated Light without which I
should not be eye',[15] and unless he pronounced his *fiat lux*
upon me, the abyss within me would be dark indeed. He is the
hearth from which the souls of men, like so many lamps, take
their light.[16] He is the *ipse qui illuminat* at the heart of reason.[17]

> *Lux lucis et fons luminis*
> *Diem dies illuminans.*[18]
>
> (Light of light, and source of illumination,
> Day which enlighteneth the day.)

That is to say, there is something sacred in our humble reason.[19]
    In this way the scruples of agnosticism and the self-suffi-

----

(3) Finally, because it seems better calculated to show that the knowledge of God, even natural knowledge, as it is in concrete reality, above all
when it is in some sense and in some degree knowledge of a personal God,
transcends the profane order, wherever we come across it, and leads us
into the domain of the sacred.

14. St. Thomas, *Tertia*, q.v, a. 4, ad 2m; *De Veritate*, q. xvi, a. 3.

15. Gabriel Marcel, *The Mystery of Being*, II: 'Faith and Reality' (1951),
p. 178.

16. St. Augustine, *De civitate Dei*, XI, xxvii, 2 (PL, xli, 341).

17. St. Augustine, *Soliloquies*, I, vi, 12 (PL, xxxii, 875).

18. Hymn, *Feria secunda*, at Lauds (St. Ambrose).

19. As can be seen, there is no question here, as some critics, obviously
owing to a misreading, have made us say, of revelation 'addressed to our
humble reason' (as though to supplement a need), but in the first place of
that humble reason itself, which is thus, on the contrary, magnified.

ciency of the profane are set aside. But man must have the courage to use his reason; he must neither despise the power
which is in him nor take pride in it. And in the supreme use
of his faculty of knowing, he should be neither hesitant nor
sacrilegious. Be the windings of his thought what they may—
let him at last find his way to the source and return to the spring.

Above all—though this may seem no more than a secondary
thought, very often repressed—God reveals himself continuously to man by imprinting his image upon him.[20] That divine
operation constitutes the very centre of man. That is what
makes him spirit and constitutes his reasonableness. That is
why, strictly speaking, no other revelation of God is absolutely
necessary; that 'natural revelation' suffices, quite apart from
any supernatural intervention. But in order to avoid exaggera-
tion, let us say that it suffices in principle. Sin has not entirely
extinguished it. For if the human soul only knows itself in
actual knowledge, reaching that knowledge through its acts,[21]
it possesses nevertheless a certain 'habitual knowledge' of
itself, real in spite of being obscure and veiled, constant
although for ever fleeting—owing to the fact that it is always
present to itself;[22] the presence of the soul present to itself,
in which it may learn, as in a mirror, the presence of God to
the soul.[23] In the same way that the reality of the divine image
in the soul is at the centre and principle of all rational activity,
which should lead it from knowledge of the world to the affirmation of God, so, in the same way, the soul's habitual knowledge of itself can become the principle of an intimate process

20. E.g. St. Thomas, *Prima*, q. xlv, a. 7; and St. Augustine, *De Trinitate*,
XV, viii, 14 (PL, xlii, 1067).
21. St. Thomas, *In Boetium de Trinitate*, q. i, a. 3; *De Anima*, a. 3, ad 4m.;
*Prima*, q. lxxxvii, a. i; *De Veritate*, q. viii, a. 6, etc.
22. St. Thomas, *De Veritate*, q. x, a. 8.
23. St. Thomas, *De Veritate*, q. x, a. 2, ad 5m.

of reflection enabling it to recognise its reality as 'image'.[24]

 . . . Forget, then, your greatness and confess your dependence. Reflect upon the splendour you bear within you.[25] Do not neglect the light that is given to you, but do not attribute the source to yourself.[26] Try to discover your reality as a mirror and as an image. Know yourself by knowing your God. Begin, as far as it is possible for a mortal, to contemplate his Face in recollection.[27]

24. Cf. A. Gardeil, *La structure de l'âme et l'expérience mystique* (2nd ed. 1927), vol. ii, pp. 95, 111, on the three moments of the soul's knowledge of itself and their 'dynamic interrelation'. St. Anselm, *Monologion*, lxvii (PL, clviii, 213B). St. Augustine, *Soliloquies* I, i, 4 (PL, xxxii, 871).

25. Eckhart, *Of the Kingdom of God*: 'Courage, noble soul! Reflect upon yourself, reflect on the splendour you bear within you: are you not honoured above all other creatures by your resemblance to God? Disdain all that is small, for you are created for what is great!'

26. Cf. St. John Chrysostom (PG, lx, 411–414); St. Augustine, *De spiritu et littera*, xii, n. 19 (PL, xliv, 211).

27. St. Augustine, *De Trinitate*, XV, c. xxiv, n. 44 (PL, xlii, 1091).

# I

## THE ORIGIN OF THE IDEA OF GOD

THERE are many theories to account for the origin of the idea of God, and in the course of the last century or so they have greatly increased in number. Most of them explain nothing or, without realising it, disintegrate the idea they set out to explain. The most diverse intellectual disciplines are confused one with another, and the *a priori* consideration which governs them all is that they are concerned with an illusion. At the very beginning we find atheism, which continues to guide their steps; little wonder if we find it at the end of the journey. Indeed the more or less explicit formulation of the conclusion is that the very idea of God must be rejected because we know 'the mechanism by which humanity constructs the idea, and that mechanism is an illusion'. But that is, in fact, to beg the question.

It is said that man deifies the heavens. Let it be granted. But where, exactly, did he find the idea of the divine which he applies to the heavens?[1] Why do we discover the same spontaneous movement wherever we look at our kind? Why that impulse to deify, whether it be the heavens or some other thing? Again, the word 'god', we are told by the philologist, simply means 'the luminous heaven of day'. Agreed. But why exactly should the 'luminous heaven of day' become man's

1. Cf. Mircea Eliade, *Traité d'histoire des religions* (1948), where the question is fully and carefully treated.

god? Yet there are many people who do not even perceive the question this raises.[2]

Those who maintain that one may, in the strict sense of the word, speak of the genesis of the idea of God—whether they conceive of it as ideological or sentimental, individual or social, and regardless of whether it is totally illusory or relatively well-founded—all deny, implicitly at least, the idea of God. They deny it by tracing it back to *something else*.[3]

'It is not very easy to see', M. Mircea Eliade writes, 'how the discovery that the primal laws of geometry were due to the empirical necessities of the irrigation of the Nile Delta can have any bearing on the validity or otherwise of those laws.'[4] We can argue here in the same way. For it is really no easier to understand how the fact that the first emergence of the idea of God may possibly have been provoked by a particular spectacle, or have been linked to a particular experience of a sensible nature, could affect the validity of the idea itself. In each case the problem of its birth from experience and the problem of its essence or validity are distinct. They are problems of a different order. The problems of surveying no more

2. It is by a similar sophism that evolutionist thinkers explain the gradual transition from the naturalist to the moral belief in divinity—a sophism clearly exposed by Renouvier, *Philosophie analytique de l'histoire*, vol. i (1896), p. 61.

3. As, for example, Emile Durkheim, of whom M. Merleau-Ponty very rightly remarks (in *Sens et Non-Sens*, 1948, p. 177) that he 'nominally defines religions by the sacred, then shows that the experience of the sacred coincides with the moments when the totem society reveals most cohesion, and concludes that the religious life, at least in its most elementary forms, and no doubt in its superior forms also, is simply the way in which society becomes conscious of itself'. Sigmund Freud does likewise, explaining 'the genesis of the idea of monotheism' and the 'historical and psychological conditions of its formation' (*Moses and Monotheism*); or A. H. Krappe, *La genèse des mythes*: 'Animism is the one necessary basis of theism.'

4. *Le Chamanisme* (1951), p. 239, n.

engendered geometry than the experience of storm and sky
engendered the idea of God. The important thing is to consider
the idea in itself; not the occasion of its birth, but its inner
constitution.

If the idea of God in the mind of man is real, then no fact
accessible to history or psychology or sociology, or to any
other scientific discipline, can really be its generating cause.[5]
No observable 'process' suffices to account for it. And in that
sense it has no genesis—no more than geometry, to stick to
the same example, has a genesis. That does not mean that it
cannot be inferred—quite the contrary. It means that it cannot be reduced to the result, itself deceptive, of some empirical transformation. It is quite possible that as it unfolds in
consciousness it may be dependent upon particular circumstances; it may be determined by particular circumstances or
provoked by some sign or other. A particular phenomenon
may be specially fitted to awaken the mind and act as the
initial shock. It is quite possible, for example, that 'the first
conception of the word of God, as a cosmic power', may
have come to our ancestors through the mediation of 'a

5. D. J. Garat, in his *Analyse de l'entendement humain*, wrote: 'When one
sees the vast numbers of divinities, before which the human race lived for
centuries, prostrate and trembling, born of hieroglyphic writing, one is
terrified at the power of signs.' Beneath the style and science of a particular
period, one can recognise a sophism which is continually being reproduced.
Garat shows himself the 'brilliant sophist' whom one of the men who knew
him best depicted (Sainte-Beuve, *Chateaubriand et son groupe littéraire*, 2nd
ed. 1872, I, p. 62, n.). If he had intended speaking merely of the divinities,
his explanation might be discussed on the plane of scientific observation,
without its being necessary to reject it *a priori*. But if he meant, as it seems
he did, to explain the *genesis* of the idea of divinity, it betrays an *ignoratio
elenchi*. How many other examples there have been! Wellhausen tried to
explain monotheism as the effect of language on thought. The same sophism,
resulting from the same *ignoratio elenchi*, has recently appeared in Julian
Huxley's *Religion as an objective problem*.

natural phenomenon, the storm: might it not be that the grumbling of the thunder suggested the powerful and awe-inspiring voice of God?'[6] Many other hypotheses could be formulated, some more, some less probable, some well supported, others less so, which, moreover, need not be mutually exclusive, and in fact are often complementary. It is possible, and by no means without interest, to analyse certain conditions, certain processes in the discovery of God, and this is where the historian, the ethnologist and the psychologist abound in useful observations, although they are, for the most part, too partial. But however fruitful these researches, they cannot in any case enlighten us upon the essential question. For let us not repeat the sophism of 'looking for principles in origins.'[7]

There are, indeed, ways without number which do lead to God; and there are also diverse ways, ways which are certain and universally valid, along which to provide a rational basis for the idea of God. For one can 'attain *he who is* starting from no matter what objects of which one can say that they *are*',[8] and of which one must say in the same breath that they *are not*. On a very different level from the level of empirical processes, there are indeed proofs of God. And that is why, strictly speaking, there can be no genesis of the idea of God.

*

6. M. E. Boismard, O.P., *Le Prologue de saint Jean* (1953), p. 111: 'There is a trace of these primitive conceptions, in a more or less poetic form, in Psalm 29, which is a hymn to the Lord of Tempests. . . .'

7. Cf. Jules Lachelier, *Vocabulaire philosophique*, 'Origin: one can only use the word "origin" of a beginning in time, a first act. . . . A metaphysical explanation . . . should not be designated by that word . . . one should say "principle".'

8. E. Gilson, *Le Thomisme* (5th ed. 1948), p. 122; cf. English trans., *The Christian Philosophy of St. Thomas Aquinas*, p. 83.

'If we consider things in their genesis we obtain a perfect understanding of them.'[9] St. Thomas's words apply here with quite special force. For—and this St. Thomas shows very clearly—it is impossible to have such intelligence of God. The idea of God can neither be explained as an illusion whose causes have been fully understood nor as a construction of the mind.

*

There has been much discussion recently as to whether the objective affirmation of God belongs to 'logical thought' or 'mythical thought'—by which people often mean: to the realm of reason or imagination; or, as might also be said, to truth or to the world of illusion. Perhaps insufficient attention has been paid to the fact that logic, too, has its illusions, and that it is tempted to extrapolate in the realm of the imaginary; and that logic may become too 'reasonable' to find in truth him who is above reason. . . . Is the God of rationalism 'the true God'? Is its idea of God really solid and rational?

In reality the authentic affirmation of God—which is something much more than an affirmation—belongs in the first instance to the deepest operation of thought, which is not itself either 'mythical' or 'logical', although it is normally obliged to borrow the procedures of logic in order to express itself, and makes use of imagination to give itself body, in such a way that its spontaneous constructions reveal a structure analogous to the structures of myths. Perhaps, if we are to take all these elements into account, we should do better to describe it with a word of which modern abuse ought not to be allowed to deprive us, namely a 'symbolic'[10] affirmation

9. St. Thomas, *In Polit.*, I, 2.
10. The words 'symbol' and 'symbolic' have only recently been so misused in an anti-intellectual and anti-realist sense that one hesitates sometimes to reintroduce them into the language of traditional thought, where they

or even, to use another and older term beloved of the Fathers, 'anagogical'.[11]

\*

All attempts to find a 'genesis' for the idea of God—like the attempts to 'reduce' it to something else by explaining its genesis—err in some respect or other. The idea of God is a unique idea, distinct from all others, and it cannot be fitted into any system. It strikes down like a flash of lightning, and can be seen cutting through the history of humanity; it plays havoc with the laborious syntheses of ethnologists and historians, and upsets the evolutionary schemes and the erudite 'physiologies of religion'. Once the intelligence reaches maturity, the idea of God germinates spontaneously.

But however indestructible it may be from then on, it does not immediately shine forth in all its brilliance. And equally it is very far from being so fully and peacefully established in the mind that it holds undisturbed sway. On the contrary, one is tempted to think that, like the seed sown in the Gospel, it fell among thorns and thistles and was quickly stifled by the incredible proliferation of myths. Or else, if it bears fruit, it seems as though the fruit were so closely intertwined with the luxuriant vegetation of the wild seed that there was no longer any way in practice of removing the latter without uprooting the former. In so far as religion co-exists in this way with a myth, it lends the latter a power of seduction which is, in the end, turned against religion itself. The gods thus secretly nourished by the idea of God are parasites and prevent

nevertheless belong. Cf. A. D. Sertillanges, *Les grandes thèses de la philosophie thomiste* (1928), p. 80 (English trans., *Foundations of Thomistic Philosophy*, p. 89); and in the opposite sense, Ch. de Moré-Pontgibaud, *Sur l'analogie des noms divins*, in *Recherches de science religieuse* (1954), p. 344, n. 17.

11. What M. Jacques Maritain calls, in *Les degrés du savoir*, 'l'intellection ana-noétique'; see *The Degrees of Knowledge* (English trans., 1937), p. 277.

the true God from emerging. . . . Hence the 'deluge of idolatry'[12] that covers the face of the earth. So that in order to
attain to religion pure and undefiled, it seems as though we
had to sacrifice all these gods, instead of purifying them or
testing their worth, and retaining those whose claims prove
authentic. Man frees himself from superstition through
atheism[13]. . . only to fall once again into superstition. Or else
. . . or else. . . . There seems no limit to the hypotheses man
can construct. How, then, can he break out of the circle?
Whichever way he takes, his reason is faced with difficulties
and with a thousand illusions to overcome before he can
emerge triumphant! And in most cases how great is the un-
certainty and how many are the errors! And even in the cry
of monotheism itself—so rationally well-founded after all—
one can often detect a certain lack of assurance:

'O stay of the earth, and you who are enthroned above it,
whoever you are, thought can approach you only with
difficulty. Whether you are Zeus, or the Supreme Necessity,
or the human spirit. . .'.[14]

. . . Unless, that is, God himself intervenes to break the
fatal circle and elects a trusted servant who is charged with
the task of announcing him to his brethren.[15] Which, as the
author of the Epistle to the Hebrews tells us, may happen 'at
sundry times and in diverse manners'.[16]

✶

12. Bossuet, *Elévations sur les Mystères*, 7e semaine.

13. Is that not, in practice, the case with Buddhism? Its founder, no doubt,
does not deny the gods, but declares them incapable of saving man; they
themselves need to be saved, and the Buddhas are above even the greatest
of the gods.

14. Euripides, *The Trojan Women*, 884–886.

15. Cf. Council of the Vatican, *Dei Filius*, cap. ii: *De Revelatione* (Acta
Concilii Vaticani, col. 250).

16. *Hebrews*, 1 : 1.

It is only natural that the idea of God should be, at one and the same time, ready to emerge and yet menaced with suffocation; for mankind—made in the image of God, though sinful —while destined to grope its way slowly up, is nevertheless obsessed from the first moment of its awakening by a call from above. From the very beginning two tendencies have been at work to retard and deflect man's natural impulse towards his creator. One of these tendencies results from the very conditions in which the intelligence is obliged to work in order to triumph, little by little, over the darkness[17]; the other, according to the teaching of the Catholic faith, is the immediate fruit of an original moral deviation. Both tendencies, the natural and the perverse, supplement and reinforce one another in obstructing the royal road of the mind, and tempt it aside among the myriad labyrinthine paths of magic and myth. There is the tendency to confuse the Author of Nature with the Nature through which he reveals himself obscurely, whose characteristics we cannot help employing in order to think of him; and there is the tendency to forsake an exacting and all too incorruptible God in favour of something inferior or fictitious. When these tendencies work together, the categories become lifeless and rigid.[18] The world itself becomes more dense. And what should have been a sign becomes a screen. The initial vision is dissipated almost before it is perceived . . . and the divine star disappears behind its 'gross shadow':

17. Cf. Mgr. Charles Journet in *Nova et Vetera* (1950), p. 192. 'It was natural that man should pass from a mythical or magical mentality, where imagination is in the foreground, to a rational or scientific mentality, where reason is in the foreground; in other words, the night of reason had to precede its day.' See also J. Maritain, *Signe et Symbole*, in *Quatre essais sur l'esprit dans sa condition charnelle* (1939).

18. On the ambiguity of heaven in ancient philosophies, see A. J. Festugière, O.P., *Le dieu cosmique*, pp. 120–250.

*Le feu, le vent, l'air subtil,*
*La voute étoilée, l'onde impétueuse ou les flambeaux du ciel*
*Sont regardés comme les maîtres du monde.*

(Fire, wind and the subtle air,
The starry vault, the impetuous swell or the torches in the sky
Are regarded as the masters of the world.)

In the furthest recesses of our consciousness, 'the glory of
the incorruptible God', before it has had time to shine with
all its brilliance, is exchanged for the gods of nothingness or
untruth. . . .[19] Or at least the God who is really close to us
has been put at a distance[20]—to remain for a long while the
unknown God. Even to those who still preserve a memory of
him, he becomes a forsaken God. And so it becomes necessary to rediscover him by stages, groping and fumbling and
sometimes thinking that we have lost him. Even at times when
the knowledge of God seems to have made decisive progress,
he is still easily conceived of as an individual with human
passions, or, on the other hand, as a vague and diffused Force.
When we think we have exhausted the idea of God, it is no
more than a sort of *materia prima*, a being as indeterminate and
as close to nothingness as empty space; or else it becomes a
principle wholly lacking in inwardness, an abstraction with
no power of irradiation. Each new formula seems more discouraging than the last and, by reaction, provokes its contrary.
The spiritual gain is never definitive, though this alone could
stabilise and nourish the intellectual gain. The better is transformed into the worse, and the great force for good in human

19. *Wisdom*, 13 : 2; *Rom.* 1 : 23.
20. Cf. Mircea Eliade in *Témoignages*, xxviii (1951), pp. 22–26: 'If there
is one *constant* in the religious history of mankind in antiquity, it is . . . the
*banishing of the supreme God*.' This banishing of the divine, in fact, translates
the progressive fall of man into 'concrete religions' which forbid him any
form of transcendence.

affairs is enslaved to profane ends: once again man deifies his needs, his interests, his passions, his ignorance and his follies. . . . And then what should have been progress takes the form of negation. More often than not, the gods of fable are supplanted by the Divine instead of by the Living God. Religions and morals close in mortal combat. For man's inwardness is the fruit of his victory over the gods. . . . But from time to time, nevertheless, a ray of pure light filters through. The pagans themselves have their 'hidden saints', and the true God chooses his prophets where he will.[21]

✳

There are many facts which make the Marxist theory, and theories like it, plausible. The whole religious system varies and presents different characteristics according to whether man is a hunter, a shepherd, or cultivates the soil. And among unbelievers Marxists are not alone in emphasising this sort of law. Research as a whole confirms it, and the historical-cultural school have made it the principle of religious evolution and have applied it rigorously to all forms of religion, apart from supernatural revelation. They have classified the religion of 'pastoral' peoples, of those who 'gather fruits', of the 'hunters' and of the 'planters'. . . . Similarly, it has been observed that in a civilization where the horse dominates the economy, the gods adored are chivalric, etc. It is also a fact that the gods of small states, bounded by narrow frontiers and centred upon themselves, are unlike the gods of the great cosmopolitan cities. As the social group expands from tribe to city and then from nation to empire, the cosmic consciousness is ordered and organised accordingly, and involves a

21. See Jean Daniélou, *Les saints 'païens' de l'Ancien Testament* (1956); English translation, *Holy Pagans of the Old Testament* (1957).

series of parallel transformations in rite and myth.[22] It is perfectly true, therefore, that myths and rites reveal and mirror social conditions—which in their turn are closely dependent upon economic conditions—and in consequence these religions tend to lend their constraints to reinforce economic conditions. Yet, to be perfectly just, it should be noted that, for all its social abuses, religion thus envisaged consecrates the very principle of society; thanks to the social and mental coherence which it ensures, it contributes more than any other element to enabling man to live, which is the prime condition of progress.

There is, however, something else to be taken into account: the essential point. One might perhaps say that Marxism, like rationalism, is quantitatively right—somewhat in the same way that determinism is true in respect of the greater part of human action, at least as far as appearance is concerned. Historical materialism is one of those basic truths which cannot fail to convince at first sight, but which is of no help to those who desire to penetrate to the heart of the real. Where experience is concerned, do not the false and the insignificant attract infinitely more attention than the substantial and authentic? The fakes and illusions of the mind, its habitually lazy or bastard forms, its repeated failures, its standardised products, like its sudden unforeseen errors, are all plainly visible, and the observer cannot fail to see them. The area they cover is vast; they encumber the scene. Whereas the thing that counts most, the first sign of change, and the seed of things to come, is almost always rare and hidden, though its action may be widely diffused and may permeate everything. But even if it happens to be noticed, it still needs to be envisaged from within if it is to be appreciated at its proper value; by a

22. This had already been remarked upon by Benjamin Constant, *De la Religion* (1825), II, pp. 2, 6–7.

method, that is, which has nothing to do with statistical methods and is beyond the scope of empirical observation. There is every reason to think, for example, that the Marxist analysis, applied conscientiously and as intelligently as possible twenty centuries ago, in Palestine, would have overlooked the humble fact summed up in a name: Jesus of Nazareth—as in fact the Jewish and Roman historians overlooked it. That almost imperceptible fact slipped through their nets, and if it happens to be caught in the mesh of learned explanations, it is emptied of its explosive force.

Nevertheless, there are certain broad lines which are too prominent to remain entirely concealed from any who will simply open their eyes. We are told, for example, that the cult of a God without form mirrors an age of trade with distant parts, and a banking system. Is monotheism the result, then, of the slow unification of the powers of the earth?—How are we, then, to explain the history of India, where profound systems of religious philosophy and exalted forms of adoration blossomed in a primitive economy and a politically amorphous society? And, above all, have people read the precepts of the Jewish Decalogue? (Their precise date is not, in this context, of great importance.) 'Hear, O Israel! . . . I am the Lord thy God. . . . Thou shalt have no other gods before me. Thou shalt not make thee any graven image. . . . For I, the Lord thy God, am a jealous God.'[23]

*

It does not require any special powers of observation to distinguish two kinds of 'monotheistic' religion in our Western world, in spite of their multiple implications and their diverse origins. The first is, at least partly, the fruit of social and political development as well as of reflection. Little

23. *Deut.* 5: 1–9.

by little the pantheons are formed in the image of what happens on earth. The gods are organised, a hierarchy is formed, and their very number and variety begin to suggest the unity of the divine. In the end the head of the divine society grows into the supreme god, while the remaining gods are no more than his manifestations or his serfs. As a nation comes to know the gods of subject nations, it amalgamates them with its own by understanding them as equivalent in a way that is at once enriching and unifying. If, perchance, there is competition, the gods of the vanquished, themselves defeated, are eliminated, unless, indeed, they are adopted by the conquerors or become demons. . . . Such, with a hundred variations, is the case in Babylon and in Egypt, among the ancient Indo-Europeans, and in the Achaemenid Empire, as in the Hellenistic world and in Rome under the Empire. . . . Can the result be said to have been all gain, politically, culturally and for thought? For the most part, yes, and sometimes the profit was very considerable. But can it be said to mark religious progress, properly speaking? Not always, and sometimes not at all. Even in cases where anthropomorphism was transcended, the goal reached was hardly more than an abstract Divine or divinised Nature: *Aequum est, quidquid omnes colunt, unum putari: eadem spectamus astra, commune coelum est, idem nos mundus involvit. . . .*[24] (It is right that whatever everybody worships should be considered one thing: we behold the same stars, the heavens are common to us all, the same world embraces us.) The concentration of gods has not given birth to God!

In the second type of monotheism, on the contrary, the one God affirms his uniqueness with a fierce, exclusive jealousy. 'There is no God but God.' He is the result neither of concentration nor of syncretism, whether political or intel-

24. Symmachus, *Relatio*, n. 10.

lectual. He imposes and sanctions a new order of values. He is the God who cannot be reached by way of the gods; the path to him leads through conversion and the breaking of idols—those made with hands and those fashioned in the heart. He is the God who throws down his gauntlet to the gods of Nature—just as the unknown young David threw down his challenge to the celebrated giant Goliath. A God who must be followed, though it involves leaving the country of one's fathers behind one . . . A God who leads into the unknown. A God who scandalizes those he does not attract. And before his face 'the gods of the nations' are nought but 'wood and stone': vanity, nothingness, 'the abomination', 'sin,' 'filthy' and 'impure', 'corpses'. They are the 'non-gods'. 'Do not forsake the Lord and adore the gods of nothingness.'[25] 'Behold, the Lord will come on a white cloud and all the works of the Egyptians will be swept away before his face.'[26] 'A jealous and exclusive God, who divides everything and leaves nothing standing before him.' Just now we were dealing with an easy-going Principle which justified the practices of polytheism and consolidated the dominions of the flesh, while remaining in itself the possession of a small élite of the wise. Now we find a Being in no sense abstract, though purely spiritual; a living Being who, although invisible, acts; an intransigent Being who demands that all worship should be addressed to him, and who wishes to be recognised by all men; a transcendent Being, though none the less powerfully personal, who overflows the boundaries of all the cities of the earth; not a cosmopolitan God, but a God who is to be, if he is not already, the universal God.

The second form of monotheism alone is charged with power. It alone is pregnant with religious progress, for it is

25. *The Book of the Secrets of Enoch*, ii.
26. *Isaiah* 19:1.

the source of a radical metamorphosis in the theory and practice of religion. It alone can promote moral and social progress,
even where it does not actually initiate it. The God of this
monotheism is the only God who can be the object of faith
in the full sense of the word. In its encounter with the first
form of monotheism, the second form makes no attempt to
compromise or to compose the differences between them;
it must, in the first place, be victorious over it. *Hebraeorum
Deus a Romanis non receptus, quia se solum coli voluerit.*[27] (The
God of the Hebrews was not accepted by the Romans, because
he wished that only he should be worshipped.)

Then, perhaps, it may utilise the first form of monotheism
to express and complete itself or in order to spread abroad, in
that way leading the first form to its goal. Now we do not
find the second form of monotheism appearing in the great
unified states at the conclusion of periods of expansion and
conquest, any more than at the end of periods of profound
speculation, or in the wake of economic changes. As far as
can be seen from the hopeless state of the sources, the religion
of Zoroaster, 'the least pagan of pagan religions', a religion
whose divine forces are not gods so much as the 'attributes
of the unique divinity',[28] arose in one of the most remote
provinces of Iran, far from the focus of culture which existed
at that time in Babylon, and before the era of syncretism was
opened by the conquests of Cyrus. The history of Judaism
and of Islam, too, belies all the theories of the development
of religion, which only invoke factors which are foreign to it.
Israel was a small nation with a rudimentary economy, a crude
philosophy and a civilisation far less developed than that of its

27. St. Augustine, *De consensu evangelistarum*, I, xviii, 26 (PL, xxxiv,
1053–54).
28. Jean Pierre de Menasce, O.P., *Le monde moral iranien* in *Les morales
non-chrétiennes* (1954), p. 42.

great neighbours, who each in turn crushed it. If Israel was quick to profit by their wider conceptions, especially during the exile, it did so for its own ends, and in order to clothe the God which it alone affirmed in more magnificent array. Moreover, it was during the Captivity, in a period of ruin and defeat, that the Israelites celebrated the triumph of their God. As for the Arabs before the Hegira, they can hardly be said to have been united. The idea of God in its highest manifestations and in its humblest forms always bursts and overflows social as well as mental categories. One may indeed say 'the Spirit bloweth where it listeth'.

\*

A lasting religion must have roots, and its birth depends upon a series of conditions which are not all of them of a religious order. No Christian need be astonished at the fact, for he knows the place occupied in religion, even in its revealed form, by the idea of 'the fullness of time'. Supernatural is not equivalent to superficial. The divine does not exclude the human. Nor is it arbitrarily superimposed upon it. But in this sphere, once again let us be on our guard against mistaking conditions for causes.

\*

In paganism, the progress of reflection tended towards the elimination of the gods. Through Christianity, faith in God has promoted the development of consciousness. Man, called by God, has come to know himself by learning to know his vocation. He has become a person for ever, to himself.

\*

If *God* is given the same name as the *gods*, it is not because of some parentage, however remote: as though, for example, the one were the perfection, the sublimation or the unification of the many. It is in order to mark the fact that the others never had but a borrowed, or rather a stolen, existence. God comes into his own, into the rights which, ever since the day when man turned away from him, have been usurped by idle phantoms or by the forces of evil.

*

There are some who think that the one God is the product of a religious *evolution*. Scattered at first in a dust-cloud of sacred beings, the divine slowly takes form; a hierarchy is organised, and by a gradual process of concentration is ultimately raised into a supreme divinity, while all the other powers created by the mythical imagination become from thenceforward its servants. Then, at leisure, it is purified, spiritualised and refined—perhaps to vanishing-point.

There are others, on the contrary, who hold that the one God is posited at a single stroke by a religious *revolution*. He affirms his position instantaneously in opposition to all else. An individual God who rejects the other gods. A certain conception of the divine which springs up in all its exclusiveness, in opposition to the conceptions entertained up to that time, when man has tired of them either because he no longer perceives their value or because he recognises their emptiness.

Both theses are based upon careful observation, and each of them deserves to be given full weight, though if one considers the living God of religion rather than the supreme principle of philosophy, there is more historical truth in the second theory than in the first.[29] The God of the Bible is

29. Cf. Edgar Quinet, *Génie des Religions* (*Œuvres Complètes*, I, 273 f.): 'Correct, embellish and complete Baal and Astarte as you will, you will never make the God of Moses out of the gods of Canaan.'

named: he is Yahweh, and affirms his uniqueness by raising up and forming his own people, distinct from all others, by imposing a particular legislation upon them, and through his Prophets he makes a mockery of the gods made with hands. The God of the Gospels is no less personal: he is the heavenly Father, and Christians can only look upon the gods of paganism, if they treat them as having any real existence whatsoever, as demons. Nevertheless, it is true that the formative, intransigent phase, during which monotheism or monolatry is established, is closely followed by an enveloping movement, a phase during which, without allowing himself to be in any way contaminated, the victorious God takes over and uses to his advantage all that is true in the thought and worship that had gone astray. The phase of opposition is succeeded by a phase of absorption, so that the two theses would appear to be complementary rather than contradictory.

Yet neither theory goes to the root of the matter. Neither goes back to the source. In reality, the idea of the one and transcendent God does not arise historically as the result of criticism or in the wake of disillusionment. It is not the fruit of an immanent dialectic whether revolutionary or evolutionary. Nor is it obtained as the result of a synthesis, as though in answer to the need and impulse to unify the fragmentary expressions of the divine; nor is it an antithesis, as though man had become conscious at last of the vanity of his age-old gods. Neither integration nor contrast can explain it. What we take to be cause is in truth effect. The idea of the one God springs up spontaneously at the heart of consciousness, whether as a result of the exigencies of reason or of some supernatural illumination, and imposes itself upon the mind of itself, of its own necessity. In fact, the clearest instance shows God revealing himself, and in doing so dissipating the idols or

compelling the man to whom he reveals himself to tear them
from his heart: *Reverberasti infirmitatem aspectus mei, radians in
me vehementer et contremui amore et horrore*[30] (You have beaten
upon the weakness of my sight, shining upon me with power,
and I shook with love and dread). First comes the 'radiation',
while the light and the attraction, interwoven with fear,
emanate from it. The phenomenon of 'reverberation' exposes
the infirmity of human conceptions, and sets them in the full
light of day, and the man whom God has touched is filled with
horror at the thought of the phantoms which he had engendered. The faith which is born in him liberates him from
superstition.

At the very beginning then, there is an encounter, a contact, a certain apperception, or whatever term may, according to the case, be applied to it—an illumination of the intellect, vision, hearing, faith. The antithesis comes second, and
the synthesis, in so far as one may use the expression, comes
last.

The first moment alone, in fact, counts. There is Abraham,
hearing the call which tears him from his country and his
ancestral cult; and Moses, receiving the Law on Sinai; Isaiah,
contemplating the majesty of Yahweh in the Temple. . . .
Jesus, moved by the Spirit and conversing with the Father.
In none of these cases is there any suggestion of a 'dialectic',
of the swing of the pendulum, the alternation of 'for' and
'against'—no trace of relativity. All forms of dialectic,
whether historical or not, and whatever their mode, imply
contrariety and negation. Whatever the spring of the dialectic,
one term is always called forth by another. The swing of the
pendulum does not imply the introduction of a new principle.
Dialectic is a powerful weapon because it corresponds to one
of the essential processes of the mind. But when it tries to

30. St. Augustine, *Confessions*, VII, x, 16 (PL, xxxii, 742).

engender thought, instead of organising it, its soul is a blind necessity. It throws no light on the inwardness of the beings which at each step, and turn by turn, it posits; or rather those beings have no inwardness; they are terms which are wholly relative to those with which they form a series. Once the idea of the living God has fallen like a seed into conscious-ness—whether by the light of reason or as the result of a supernatural revelation—it certainly is subjected, like all other ideas, to dialectic. In a sense more than any other idea, since it becomes the principle of perpetual 'ferment' which works unceasingly in it. But none the less, it remains substantial and positive, and that is what ensures its victory. Far from corre-sponding to a phase in human dialectic, it is, on the contrary, dialectic which plays the intermediary role, linking in its process a reality already perceived and a mystery surmised, without ceasing to be sustained in its movement by a pre-sence. . . .

Observe, again, how much more striking it appears in the concrete dialectic of history. Religious monotheism, even as we owe it to Israel and Christ—and this is true in some small measure in certain analogous cases—is illuminated by the divine source. Before being a belief, and *a fortiori*, before becoming a tradition or an idea, it was a vocation, and remains one as long as it preserves its authentic vitality. Its formation bears no trace of the dialectical movement of *ressentiment* in the Nietzschean sense. Abraham did not find the true God by turning against the gods of his ancestors; he had to struggle in his heart to abandon them: his faith had to be a victory. Jesus did not preach the vanity of this world like the Buddha, or the vanity of the gods who canonise it, because they are the mythical form of its very substance: he proclaimed the King-dom of Heaven in which his soul already breathes, and he reveals the Heavenly Father's love in and through his own

person. In that sense, too, the Apostle's words are verified:
There is only Yes in him.[31]

＊

Mythologies have been psychoanalysed with more or less
success. More and more it becomes necessary to psychoanalyse atheism. Attempts to psychoanalyse faith will always
end in failure.

---

31. 2 *Cor.* 1 : 19. Cf. Karl Barth, *Dogmatics in Outline* (English trans., 1949),
p. 40. 'Once the true God has been seen, the gods collapse into dust and
He remains the only One.' And for a comparison with the Buddha, see my
*Aspects du Bouddhisme* (1951), I, 51–53.

# 2

## THE AFFIRMATION OF GOD

IF, like St. Augustine, we only use the word belief to denote
the acts in which the spirit adheres to truths beyond the grasp
of the senses, and which the intelligence cannot as yet penetrate, we may say that the affirmation of God is always an
act of faith. But then we must be more precise, and add at
once that no other affirmation can compare with it in certitude.
For even before it was formulated, before God was named,
that belief provided the foundation of all others. All affirmations, as Descartes rightly saw, though he explained it badly,
depend upon that belief, and certainty in all its forms is
rooted in it. 'However we set about it', Leibnitz, for his part,
declares, 'we cannot do without the divine existence.' The
two philosophers, each in his own way—which it is not our
present concern to criticise—had taken up the axiom enun-
ciated by St. Thomas Aquinas: 'All knowers know God implicitly in all they know.'[1]

1. *De Veritate*, q. xxii, a. 2, ad 1m.; cf. Joseph Maréchal: *Le point de
départ de la métaphysique*, cahier V (1926), p. 337; and Hans Urs von
Balthasar: *Phénoménologie de la vérité* (tr. R. Givore, 1952), p. 35.

It has been said: 'It is obvious that the thought of St. Thomas is diametri-
cally opposed to that of Descartes and of Leibnitz, to which it has been
likened.' Let us say, with more moderation and truth, that the thought of
those two philosophers (who differ from one another) is manifestly different
from that of St. Thomas, to which we had no intention of likening it. But
those differences only make the very real analogy, at the point which we
have indicated, all the more remarkable.

Every human act, whether it is an act of knowledge or an act of the will, rests secretly upon God, by attributing meaning and solidity to the real upon which it is exercised. For God is the Absolute; and nothing can be thought without positing the Absolute in relating it to that Absolute; nothing can be willed without tending towards the Absolute, nor valued unless weighed in terms of the Absolute.[2]

So it is not only in the acts which we call religious, nor in a crudely pragmatic sense, that *God is used*—to recall a well-known expression. The supreme contradiction is to use God in order to control the flux of existence, to organise chaos, to make statements, to judge, to choose—in a word to act spiritually and not to fall into contradiction at each step, and then, simultaneously, refuse to recognise him; to think him away without whom thought would only be a psychical manifestation: the supreme contradiction is to lean upon God in the very act of denying him. That, indeed, is a judgment which denies and destroys itself, not only where its content is concerned, in itself, but by undermining its own structure and refusing the condition of its existence. No doubt the contradiction remains unseen, because it does not intervene between two objective affirmations, but between objective and transcendental affirmations; the contradiction is between the assertion expressed in words and the assertion lived by thought. It is not, consequently, a particular or logical contradiction—which is why it is always possible—but a total, vital, spiritual contradiction—a contradiction in the being who thinks, a sin by the spirit against the spirit.

That is how the pagans behaved when they sought sanctuary from the barbarians in Christian churches and took advantage of the security offered them by the God of the Christian in

2. Cf. Maurice Blondel's answer to L. Brunschwicg in *Bulletin de la Société française de philosophie*, 24 March, 1928, p. 53.

order to blaspheme his name. One would have to stop willing and thinking to have the right to deny God without contradicting oneself.[3] One would have to abandon speech.[4]

\*

One cannot sever the mind's relation to the Absolute—the Absolute thought as real—without destroying the mind itself. One cannot do away with 'that primary relation to being which the philosophers of progress and the philosophers of totality invariably ignore'. Unless we are to close the door to all philosophy worthy of the name, we cannot refuse to acknowledge that 'basic experience'—the presence of non-conceptual being to consciousness which is common to the philosopher and to all men.[5]

\*

'God, being without principle, cannot be affirmed by virtue of a principle distinct from himself' (P. Scheuer).

That does not mean that reasoning—according to logical principles—in order to prove the existence of God is superfluous, but that the thought, which is our affirmation of God, is not the conclusion of an argument. The thought which reasons comes before the reasoning. And if reasoning must perforce intervene, that is in order to show us what thought consists in or what it implies. For the reality of thought is not a fact in the psychological order, and cannot be explored by empirical observation, whatever the method employed.

This comes down to saying, once again, that the existence of God is not a truth among other truths, a particular truth

3. Cf. Aimé Forest, *Du consentement à l'être* (1936), p. 104.
4. Cf. André Bremond, S.J., *Une dialectique thomiste du retour à Dieu* (*Nouvelle Revue théologique*, 1935, p. 569).
5. Ferdinand Alquié, *La nostalgie de l'être* (1950), pp. 144, 148.

dependent *in itself*, upon another, greater truth, or a more comprehensive and fundamental truth of which it is, in some sort, one of several possible applications. In other words, the Being of God is not a particular Being with its place among others, at the beginning of, or within, a series. God is not the first link in the chain of being.

Or again, we must recognise—in face of rationalism in all its forms, and all forms of contempt for the certainties of reason—that God is the reality which envelops, dominates and measures our thought, and not the reverse. He is the reality which makes our thought at once so great and sure of itself, so absolute in its judgments and so necessarily obedient.

In short, the reality of God must be taken seriously, and the transcendence of God acknowledged with all that it implies.

✳

If we did not have a certain idea of God—not yet seen, not objectified, not conscious, yet present to consciousness, and in fine, *not conceived*—previous to all our concepts and always present in all of them, the purification to which we subject them in order to think God correctly would ultimately serve no other purpose but the denial of everything and a final nihilism. To speak of the phases which succeed the phase of negation as 'excellent' or 'eminent' would then be a poor joke. For the phase of negation once conscientiously traversed would have left the mind a *tabula rasa*: it would have left nothing standing. All the terms formed with the prefix *sur* would be parrot-talk, pure logomachy, or a disguised return to the original affirmation such as it was before the critique was instituted.[6]

6. This danger can never be removed once and for all. St. Thomas Aquinas was conscious of it. For him, too, 'the analyses in which negation triumphs, less favourable to illusion than superlatives, unfold in an atmosphere of

In the same way, if there were no idea of God whatsoever prior to the reasoning by which we try to provide him with a logical basis in our thought, the critique which we must necessarily make of the general form in which those arguments are set would terminate in the denial of the affirmation of God.

But the proposed hypothesis is worthless. The dual activity of the mind is not like the work of Penelope. It really does reach a conclusion. Its success is definitive. That is because the idea of God is mysteriously present in us from the beginning, prior to our concepts, although beyond our grasp without their help, and prior to all our argumentation, in spite of being logically unjustifiable without them: it is the inspiration, the motive power and justification of them all. *Omnia cognoscentia cognoscunt implicite Deum in quolibet cognito*[7] (All knowers know God implicitly in all they know).

Such an assertion is valid of God alone. However moderately one interprets it, it is priceless. The idea of God presides over our negations and our critiques, rather in the same way that a word on the tip of one's tongue—a word one knows perfectly well, tries to recall and cannot quite articulate—will brush aside and eliminate all the other words that present themselves to the mind. The affirmation is triumphant in the midst of our negations, and the critique itself is a confirmation.

In its primary and permanent state the idea of God is not, then, a product of the intelligence. It is not a concept. It is a

mystery.' M. D. Chenu, O.P., *Introduction à l'étude de saint Thomas d'Aquin* (1950), p. 140. *Deus, qui scitur melius nesciendo*: as M. Etienne Gilson says (in the *Bulletin* quoted in note 2 above), that is 'a classical form of Thomism'.

7. St. Thomas's principle is equally that of Duns Scotus. Because of that principle and of all that corresponds to it in Thomism, it has been said that 'taken in a new sense'—the sense which we are trying to define—'exemplarism is one of the essential elements of St. Thomas's system' (Gilson, *op. cit.*, p. 197).

reality: the very soul of the soul; a spiritual image of the Divinity, an 'eikon'.[8]

✳

If the mind did not affirm God—if it were not the affirmation of God—it could affirm nothing whatsoever. It would be without laws; like a world deprived of its sun. It could no longer exercise any rational activity, and could only sink back into the dark limbo of obscure psychical subjectivism. It could no longer judge. It would have lost its light, its norm, its justification, its point of reference, the one thing which can serve as a foundation for all else.

That does not, of course, mean that we do not have to prove God. It does not mean that the existence of God is obvious from the word 'go'. Reason rises to the Absolute from the relative. And what supports and directs its steps is, from another angle, the goal of its journey. What explains and justifies knowledge must be established by knowledge. And in order that we should recognise it, that which is at the base must appear at the summit.

✳

Where belief in God is concerned, I cannot rest content with a doubtful argument, and an inconclusive proof is as repugnant to my moral sense as it is offensive to my intelligence. The importance of the matter at stake is no reason to be easy-going; on the contrary, it obliges me to be more strict. But on the other hand, if I were more clear-sighted, a

8. The tremendous importance of the notion of the image of God imprinted in man had not escaped the Fathers of the Vatican Council. There are two passages, from the pen of Mgr. Gasser, which testify to it. *Acta et Decreta SS. Concilii Vaticani (collectio Lacensis,* t. vii, Freiburg im Breisgau, 1890), coll. 132, 149.

mere suggestion, a mere hint would suffice: for in fact I bear the proof within me. Before expressing it with a greater or lesser degree of learning and critical acumen, I am conscious of its role as fulcrum. I raise my mind to God as I breathe; in each case by virtue of the same necessity. There is, however, this twofold difference: where the body is concerned there is no escaping the necessity to breathe; whereas in the case of the spiritual life, it brings its own light with it; whereas, by an incredible paradox, while the darkness in which the body is enveloped does not prevent it from breathing with perfect regularity, the light which accompanies the respiration of the spirit is not sufficient to make it recognisable, and the mind can, without doing away with it altogether—which would mean death—at least disturb and trouble its breathing.

*

If, to strengthen my belief in God, I find it necessary to have recourse to external means, that is not because my intellectual certitude is in the smallest degree unsettled. If the objection put to me is valid at all, then I know that it is only valid because I am lacking in skill. I have never really entered the labyrinth in which I am supposed to have been trapped.

And then I know, too, that the intelligence is not the whole man. An intelligent man does not regard himself as pure intelligence. All the appeals to custom, to tradition, to authority, to the positive teaching of religion, to the gestures repeated since childhood—recourse to the 'mechanism'— are not meant to compel reason nor to supplement it, but to protect it against the vertigo of the imagination. Their real purpose is to calm the child who, according to Plato, lives on in all of us. And the only people to be scandalised are, in the words of St. Augustine 'those who do not know how rare and difficult a thing it is for the fleshly imagination to be subdued by the serenity of a devout mind'.[9]

*

Metaphysical truths, however rigorously they may have been deduced, do in fact leave the door open to an element of doubt. Even those who are most forcibly struck by them 'are afraid, an hour later, that they may have been mistaken'—in any case, they are not satisfied with them. Not that these truths are established on weak foundations or that the intelligence pure and unalloyed does not confess itself convinced by the proof. But when their demonstration is over the recollection of it is not always strong enough to repel the assaults of contrary impressions. Their light may, perhaps, shine in an abstract heaven: but it is meant to be felt—*est enim sensus et mentis*[10]—not simply proved; to be possessed, embraced and not merely perceived in the dim distance, draped in a pale and superficial clarity. Now the proof imposes these truths, but it does not give them to us.[11] The certainty it confers is not given to us as a possession. It is well, indeed, that man should be able to prove his immortality to his own satisfaction. But for an immortal being, how can that proof be more than a *pis aller*? However rigorous we assume the proof to be, it is powerless to disperse the sense of unreality which accompanies it even in a clear light. The more we feel the proof as proof, the more conscious we become of the misery

9. St. Augustine, *Contra Epistolam Manichaei*, ii, 2 (PL, xlii, 174); also *De Trinitate* IV, prooemium (PL, xlii, 887).

10. St. Augustine, *Retractationes*, I, i, 2 (PL, xxxii, 585–586).

11. This does not mean, as has been wrongly supposed, 'the proof of them is not given to us', which would amount to a contradiction. It seems to us evident that the abstract proof of a real object does not give us possession of the object. If the object itself were given to us, if we really possessed it, then what need would there be for the proof?

of the human condition which obliges us to resort to it, and which remains after it has been provided.

And when we turn our minds to God, the infinitely pure Being 'who lives in inaccessible light', the consciousness of our misery is only sharpened, and that sense of unreality weighs heavier still upon the mind. How can we be satisfied with a proof where God is concerned—God who is above all essence, all names and all forms, whom nothing ultimately can represent, though all things indicate him[12]—God who because of his perfect 'actuality' and his plenitude cannot be directly conceived by objective reason except as bare existence whose reality is both intimate and impossible to grasp, and that no 'introversion' can ever yield us as a lasting possession.

*Ubi est lux inaccessibilis, aut quomodo accedam ad lucem inaccessibilem? . . . Numquam te vidi, Domine Deus meus, non novi faciem tuam!*

(Where is the inaccessible light, or how can I approach the inaccessible light? . . . I have never seen you, Lord my God, I have never known your face!)[13]

*Haec lux est inaccessibilis, et tamen proxima animae etiam plus quam ipsa sibi. Est etiam inalligabilis, et tamen summe intima.*

(This is the inaccessible light, and yet it is even closer to the soul than the soul itself. It cannot be grasped, and yet is most intimately present.)

✳

We are always dreaming of the impossible. We should like a truth that was not abstract and a reality that was not empirical; a fact that had all the characteristics of a law; a verification which would at the same time be the answer to a need; an ideal satisfaction which was also a real possession. On these terms only could we have perfect peace of mind. But in fact

12. St. Thomas, *Prima*, q. xiii, a. 2; *De Veritate*, q. ii, a. 1, ad 9m.
13. St. Anselm, *Proslogion*, i (PL, clviii, 225c).

we always swing between the two poles. The duality is insurmountable to our divided nature. Unity always slips the grasp of that compound of sense and reason. No sooner do we think we have seized it on the wing than it falls apart, and universal unity is not concrete unity. What in fact we attain is not merely, because of its inadequacy, the starting-point of a new inquiry: it is always a fresh disillusionment. *Cur non te sentit, Domine Deus, anima mea si invenit te? An non invenit, quem invenit esse lucem et veritatem . . .?*[14] (Why does my soul not perceive you, Lord God, if it finds you? Or does it not find you, whom it finds to be light and truth?)

Are we moved by a figment of our imagination? By no means. There is one instance in which the impossible is not a phantom. But God alone, operating in a realm beyond sense and beyond reason, can bring about the synthesis, though it is always partial and fugitive here below. The unique Presence shines in the night of the senses and in the night of the reason, in the night that remains dark.[15]

*

Why is it that the mind which has found God still retains, or constantly reverts to, the feeling of not having found him?

14. St. Anselm has felt this sort of constantly renewed disillusionment, which is nevertheless not discouragement, in the *Proslogion*, xiv (PL, clviii, 234D). It is not the same thing as that 'psychological discomfort which one experiences so easily when faced by the most rigorous metaphysical proofs', a discomfort of which it has been said that 'true wisdom consists in reacting against those unjustified impressions and these demands of the carnal man' (Fernand van Steenberghen in *Revue philosophique de Louvain*, vol. xlv, 1947, p. 313).

15. It should be clear enough that there is no question here of a natural intuition of God as an original apanage of the human mind. On the contrary, even mystical and supernatural gifts never attain more than a partial and fugitive anticipation. . . .

Why does that absence weigh on us even in the presence itself, however intimate it may be? Why, face to face with him who penetrates all things, why that insurmountable obstacle, that unbridgeable gap? Why always a wall or a gaping void? Why do all things, as soon as they have shown him to us, betray us by concealing him again?

The temptation is to succumb to this scandal and to despair in proportion as one had formerly thought to have found him: a temptation to deny the light, because the veil becomes opaque once again, or because we are blinded: a temptation to relax once we have made the effort which, as always, led us back to the starting-point. In the case of others, there is the opposite temptation, the easy short-cuts, the illusion of those who persuade themselves that they have only to tear aside the flimsy veil for the Presence to appear—in the belief that, provided they turn their gaze inward and fix it upon the luminous centre which illuminates all their thoughts, they will enjoy the sight of God—the illusion that it suffices to be in order to possess Being. . . . The temptation to underestimate the obstacles, to imagine that serenity is easily acquired, and to confuse the faint clarity of being with the divine light . . .

Why, O Lord, these ambiguities? Why those hesitations and oscillations of the mind? Why those contradictory motions of the soul?

. . . *Cur hoc, Domine, cur hoc? Tenebratur oculus ejus infirmitate sua, aut reverberatur fulgore tuo?—Sed certe et tenebratur in se et reverberatur a te. Utique et obscuratur sua brevitate, et obruitur tua immensitate. Vere et contrahitur angustia sua, et vincitur amplitudine tua. . . .*[16]

(Why this, O Lord, why this? Is the eye darkened by its own weakness, or blinded by your light? — Without doubt it

16. St. Anselm, *loc. cit.*

is darkened in itself and blinded by you. Indeed, it is obscured by its own littleness and overwhelmed by your immensity. Surely it is contracted by its own narrowness and overcome by your greatness.)

\*

It is sometimes said that the existence of God is 'probable'; but this really has no meaning. On what could that probability be based except, like any other probability, upon a more general certainty of the same order? But God is alone in his order, and the place which he occupies in knowledge is unique. The probability of God ought therefore to be based upon the previous certainty of his existence. You might as well say that our own existence is probable. . . .

Probability has no meaning outside the empirical realm. It has no sense except with reference to a particular object, an object, that is, which forms part of a group or class: one fact among others. Now, God does not form part of our common experience. God is not a fact, any more than he is an 'object'. The reality of God is not that of an event. Nor is God a particular case, the particular application or the realisation in a particular instance of a general truth, or of a universal principle existing prior to him. As the ancients said, 'Being is extraneous to kinds'[17]; God is unique, 'God is not in a genus.'

And so, in order to respect the mystery in which our knowledge is steeped, even where its most intimate certainties are concerned, we may say that the life of the spirit rests upon a belief, and that at its root is a certain kind of confidence,

17. St. Thomas, *De Potentia*, q. vii, a. 3; *Prima*, q. iii, a. 5; q. vi, a. 2, ad 3m.; *Contra Gentiles*, I, c. xxv and xxxii. Cf. Sertillanges, *op. cit.*, p. 67 (English trans., p. 75).

or better still an 'anticipation'.[18] It is better to acknowledge the sense of unreality which, in the condition of our terrestrial existence, even the most rigorous use of reason very often serves to strengthen. . . . But these various notions are not opposed to the notion of certainty, and have nothing whatsoever to do with the notion of probability.

＊

The probable may also mean what is 'likely'. But who could describe the Being of God as 'likely'? On the contrary, if we are content with analogies and appearances, if we only listen to ordinary reason and ordinary judgment, what could be more unlikely? What could be more baffling, from whatever angle one approaches it? No, I do not affirm the Being of God, the Being which is God, because it is 'likely', but in spite of its being unlikely, in spite of the antinomies I come up against in affirming it, in spite of the difficulties which continue to hold me back. But I affirm it, nevertheless, with complete assurance, not because I am impelled to do so by some extraneous impulse, but because rational necessity tells me that it is impossible that he should not be. Its light is indirect, and what it reveals is only negative; but for all that its power is such that it sweeps aside all 'likelihoods'. The unlikely is also the incontestable—and the latter infinitely outweighs the former.

＊

'Certainty is that deep region where thought can only keep its balance in action.'[19]

＊

18. It is the word dear to Clement of Alexandria: πρόληψις.
19. Jules Lagneau: *Fragments* (*Revue de métaphysique et de morale*, 1918, p. 169).

'Truth', Malebranche says in a magnificent phrase, 'is distant, it is not palpable, and it is not a good which we feel impelled to love. It calls, therefore, for the closest attention.' 'But', he adds, with St. Augustine, 'how can a man who is torn in all directions, struck from all sides, who is thrust back when he takes a step forward, dragged on if he steps back, and who is continually irked and ill-used, how can such a man apply himself to the matter?' For that, after all, is the condition of the spirit in the flesh. It is never—as yet—completely itself. Man can never give himself up for long to the contemplation and search for the truth without encountering obstacles.

That is why we need all the paraphernalia, the whole 'machinery' with which our belief in God is protected and strengthened—and at the same time justified. They do not exist to remedy defective proofs or supply rational certainty, but simply to allow or facilitate in some measure the 'closest attention'. The task itself continues to be necessary, because 'prejudices always return to the charge, and drive us out of positions we had already occupied unless we entrench ourselves securely and guard our positions vigilantly'. Those 'solid entrenchments' do not distort the truth, they save it. Thought, in that way, is preserved from 'mental vertigo'; it is not enslaved but freed.

\*

Freedom, the supreme prerogative of the spirit, is respected even in our most unshakeable certainties, whenever they concern an object that surpasses us. It is respected, then, in the supreme certainty, the firmest and most solidly grounded of all, of the existence of God. Indeed, it is only then that freedom exercises its full powers. For the human mind, in spite of a temporary dislocation, is not at bottom divided against itself.

Its faculties are not mutually exclusive. It is untrue that the two faculties of knowing with certainty and of willing freely can only be exercised at each other's expense, or by subjecting one to the other, as though a free certainty were a half certainty or only half free. On the contrary, they enhance one another in proportion as their object is exalted, and tend to meet in unity. They are never more closely united than in the affirmation of God.

\*

The affirmation of God in the subject who makes the affirmation is not only essentially and fundamentally free in Spinoza's sense, *qui sola ducitur ratione*, with the freedom which belongs to all spiritual activities, nor is it only free in the higher sense of being autonomous and possessing the initiative which freedom requires in order to fulfil itself.[20] It is also free in the sense that it possesses that humbler, empirical and everyday freedom which is always struggling to adapt itself to circumstances, forestalling surprises, striking root in daily life, and using if need be, without shame, all the little tricks that prudence suggests and that common sense recommends.

The first of these freedoms is the very condition of true knowledge, the freedom of the subject whose judgment cannot be 'compelled by any external cause whatsoever'.[21] The second is indispensable to the affirmation of God, the keystone of knowledge. The third may be useful or necessary at any

20. Any reasoning which progresses from premises to a conclusion implies an initiative, a free act in that reasoning itself—not to supplement it or inflect it in any way, but simply to give it the necessary impulse without which there could be no progress in thought.
21. Cf. Descartes, *Méditation quatrième*. Jean-Paul Sartre comments on the text in question in *La liberté cartésienne* (*Situations*, I, 1947) emphasising the voluntarist character of Descartes's thought.

moment in substantiating our reflective affirmation which maintains us in the truth of our own nature.

*

In affirmation, as in the object itself, in thought as in being, nothing is isolated: everything is joined by an unbreakable chain, and one link involves all the others. From a static, abstract point of view one can and must distinguish various stages, but what happens at one level does not invariably react on other operations occurring above or below. Each individual proof has its own degree of validity; each object its own degree of evidence. We cannot hope for sound method or healthy thought as long as there is no attempt to arrange questions in a series, and so avoid calling everything in question at every moment. On any hypothesis, the natural use of reason teaches us certain truths, though others remain unknown. It allows us to rise to the knowledge of the highest of all things, the truth of the existence of God, and even in the most essential matters a sinner may reason better than a saint.

But the problem of the knowledge of various truths, or even of their respective degree of certainty, is one thing— the problem of their *ontological index* is another. In the last analysis it affects everything which the mind affirms. This question is not governed, like the previous one, by the formal logic of the intelligence (which it leaves intact) but by the real logic of the concrete being. It bears, necessarily, upon the whole as such, and what it envisages is the activity of a living spirit engaged in an adventure which itself forms a whole. And in the last analysis, the meaning which the spirit imprints upon its adventure confers a corresponding coherence and solidity upon its mental universe.

So we shall not say that the spirit cannot by itself be certain of many things; nor even, which would be quite another matter, that it cannot attain to a distinctively metaphysical certitude. But, to be more precise, we must add that the character of metaphysical certainty is still provisional, and, above all, that the being upon which it bears does not as yet possess, if one may so express it, all its density. It is essential to distinguish here in order to simplify a process which is all *nuances* in the concrete, and is composed of infinite shades of meaning; there is the period which precedes the subject's refusal or acceptance of grace; and there is the subsequent period. During the first period, ontological certitude is what it is, and there is no reason to declare it illegitimate or, rather, illusory. After the refusal, those epithets take on meaning— and need to be carefully analysed. For although it may then be possible to describe the ontological certainty in question as illegitimate or illusory, that does not mean calling it illusory in itself—since the nature of the intelligence has not changed —but because from then on it is vitally contradicted.[22]

Man is a spirit created in the image of God. No degree of perversion can uproot that essential characteristic and inalienable prerogative. Man cannot alter the fact that the *image* is in him.[23] But if he tries to destroy it by every means in his

22. Truth and values are not only left 'in the air' for those who refuse God, 'they are positively deprived of foundation: Descartes was not wrong in thinking that the clear-sighted atheist would have no right to be a geometrician; since, although geometry is not in itself an immediate knowledge of God, to refuse God is to compromise it at its root by suppressing the very source and guarantee of all truth.' J. M. Le Blond, *Le chrétien devant l'athéisme actuel* in *Études* (1954), p. 299. Nevertheless, it remains true that 'truth, for him who rejects or refuses to live by it, is no longer the same as for the person who feeds upon it, but it still is; although entirely different in the one and the other case, its reign is not impugned in either case'; M. Blondel, *L'Action* (1893), p. 438.

23. St. Augustine, *De Trinitate*, X, xii; XII, vii; XIV, iv, viii (PL, xlii, 984, 1003, 1040, 1044).

power, if he deliberately goes against his vocation as spirit, he inevitably introduces contradiction not only into his intelligence—which may continue to function as before—but into his very being, setting his intelligence and his life in contradiction. And as long as that contradiction is not resolved, it deprives him in principle of the right, or rather of the possibility, of saying *it is*, and of giving those two monosyllables the full force which goes to the very root of things.

*

Far from arresting thought at the pure representation of a thing, to be followed soon after perhaps by the affirmation of the irrational, the judgment of existence gives thought its impulse, and frees it to advance by stages to a recognition of the metaphysical absolute. But far from making us forget the real, that movement of thought recalls us to it, and in a sense we only affirm God metaphysically in order to be more certain of the existence of creatures. In fact, the metaphysical explanation always comes to us as a victory over representation of the real which is purely abstract and without depth.[24]

*

The negativity of the consciousness, which ought not to be underestimated, is an obverse which requires an inverse. If the *for itself* (*pour-soi*) means separation from self, a negative power, it is because its true being, to which it aspires, has not been given to it. The capacity to say no, and to reach beyond all determination, would hardly be intelligible unless it expressed an orientation towards a higher form of being, a call to plenitude, the absence of which is in fact a distinctive mark of the consciousness. . . .

Our consciousness, certainly, is not fullness of being. It would not arise in a complete being as an unaccountable nothingness. It is in an incomplete and lower being that it expresses aspiration towards further being. It is in the experience in which the life of

24. Aimé Forest, *Du consentement à l'être*, pp. 107–108.

the spirit is inaugurated that I become inadequate to myself. The self can neither join itself nor equal itself. It is continually obliged to choose what it wishes to be, and its existence means giving itself that being in significant acts.

That transcendence, which constitutes our personal consciousness, . . . imposes upon all men the duty of having . . . a philosophy. The necessity of a philosophy and the presence of an absolute in every judgment are two ways of affirming the same necessary aspect of human consciousness. Indeed, that is not as a rule contested. The difficulties . . . begin as soon as we try to understand the nature of that absolute and the real character of the philosophy. . . .

No doubt, it will be said, the human mind cannot do without some notion of the whole; the mind needs to posit the idea of absolute truth. But should we conclude that that truth exists independently of it? That subsisting truth, it will be said, is really only an illusory projection into being of a category indispensable to the play of thought. The idea of the absolute plays the part of the scaffolding which thought uses in order to construct itself. And the scaffolding which was at first larger than the construction must subsequently be eliminated. But it is difficult to speak in these terms, for the idea of truth cannot just be added to thought in an optional way. Thought is consubstantial with truth. Thought is not constituted in itself prior to the idea of truth; it is the birth in consciousness of the need for truth. It is not a secondary or contingent aspiration of the mind; the aspiration is mind itself, which is only the capacity or function of truth. It is impossible that the absolute should not be, because my mind only exists through it. The mind denies it in a judgment which is only valid because it affirms the absolute. That by which my mind acquires its being cannot fail to exist.[25]

\*

25. Gabriel Madinier: *Conscience et Signification* (1951), pp. 62–67. And on 'the spell of negation' and the philosophies of negation, see Aimé Forest, *La vocation de l'esprit*, pp. 15–42.

If our concepts, by themselves, uncorrected by analogy, are only suited to the world of experience, then we must say as much and in the same degree of our reasoning in so far as it is merely the organisation of our concepts.

It will be said, no doubt, that duly selected and corrected, our concepts can be adapted to transcendental reality. That is true. We are indeed obliged to use them although in spite of everything they remain unworthy of so noble a usage.[26] But for that to be true it is necessary, in the first instance, for that reality to be posited; it must, *in a sense*, have been thought implicitly.

If we apply this reasoning to the subject in hand, it is only when the affirmation of God is first posited—an affirmation which is still implicit, implied in each of our judgments on existence or judgments of value, and in consequence co-extensive with our whole spiritual activity, an affirmation congenital to the mind—that we can try to rejoin our affirmation in our conscious life, turning it into logical form by way of reasoning: just as it is only when we are in possession of the idea of God contained in the implicit affirmation, that we can attempt to form some representation of it by the only way open to us: the way of concepts. That is the first, subterranean, phase of the mental life, unperceived but definitive. God must be present to the mind before any explicit reasoning or objective concept[27] is possible, and this is necessary if they

26. Cf. St. Augustine, *Sermo* 341, vii, 9 (PL, xxxix, 1498); *Contra Adimantum Manichaei discipulum*, xi (PL, xlii, 142), etc.

27. In order to avoid exaggerating or misrepresenting the sense of this paragraph, the words 'before any objective concept', (before any representation, any objective grasp) should be clearly noted. For 'the innateness of the natural light must not be confused with the innateness of its content' (E. Gilson, *La philosophie de saint Bonaventure*, 2nd ed., p. 297, n. 3). But one may try to bring out the significance, apart from any content, of this innateness of the natural light. And since the confusion is frequent, let us

are to perform their indispensable task with reference to him; he must first of all be secretly affirmed and thought.[28] Before he can be 'identified' by a conscious act, there must exist a certain 'habit of God' in the mind.[29]

If there is a truth 'towards which everything in us aspires and conspires, a truth which is lived before it is known, a truth that we can perceive with certainty even before subjecting it to the discipline of proofs and the control of concepts—because it is connatural to us—then it is, without a doubt, the knowledge of God'.[30]

✣

If we consider the affirmation of God where alone it exists in act, where alone it is really made, in the concrete intelligence which is at the same time a particular subject, in the responsible person that is, then the affirmation of God can be seen to be an act which is unlike any other. There is something

repeat that it is, nevertheless, not an objective knowledge, any more than an *appetitus naturalis* or *innatus* would be an actual desire, objective, 'elicit' and conscious.

28. Cf. Gabriel Marcel, *Du refus à l'invocation* (1940), p. 231. St. Thomas, Father Chenu tells us (*op. cit.*, p. 72), 'calls in question' the existence of God in order to prove his existence rationally, starting from the faith which he already has (which does not mean that the rational demonstration depends in any way upon the act of faith; and which is, moreover, not the same thing as the 'methodic doubt' of Descartes). The same thing may be said of Duns Scotus, *De primo rerum omnium principio*, c. i, a. 1. Quoting this last text whose 'fullness' he praises, Gilson does not represent it as unusual, but regards it on the contrary as 'the method of Christian philosophy' (*L'esprit de la philosophie médiévale* 2nd ed., 1944, pp. 51–52; cf. *The Spirit of Mediaeval Philosophy*, p. 52). Nevertheless, whatever may be the truth regarding these analogous cases, we are solely concerned with the implicit affirmation contained in the judgment.

29. H. Paissac, O.P., *Preuves de Dieu* in *Lumière et Vie*, 14 (1954), pp. 101–102.

30. J. Maréchal, in *Nouvelle revue théologique*, 1931, pp. 195, 204.

in it of the ontological argument, and something of the wager; though it is neither the one nor the other. It expresses the most luminous evidence and attests the most obscure truth.[31] Of all our acts, it is the most free and the most necessary. It is the most enduring of affirmations and the most personal of all engagements.[32]

✳

The more pure the light, the less it compels us.

31. Cf. Newman, *Apologia*: 'Of all points of faith, the being of a God is, to my own apprehension, encompassed with most difficulty, and yet borne in upon our minds with most power' (Part vii, General answer to Mr. Kingsley).

32. These oppositions derive, as the reader will have perceived, from the fact that this affirmation may be envisaged either impersonally, in the intelligence as such, or as a concrete act of the human being, as a 'decision of thought'.

# 3

## THE PROOF OF GOD

MANY people regard the existence of God as a matter of opinion; or if they consent to speak of certainty at all in this connection, they add, by way of excuse, that it is a question of feeling, an exclusively personal certainty. To us, God is the object of proof. On this point the Catholic Church has expressed herself more than once, helping the reason of those who have confidence in her to regain its self-confidence, and encouraging reason to face the danger which threatens it in our day: 'the abdication of metaphysics'. The movement which carries us to God beyond the 'visible and invisible' creation on which it rests is not just an impulse of the heart, accompanied at the most by an intellectual opinion. However personal it may be—and should be—in each one of us, that movement has a universal value. A *de-monstratio* could trace its itinerary, analyse its essential mechanism, indicate the source and distinguish its stages, which are valid for all minds.

But just as there are different kinds of objects, so there are different kinds of proof. Provided a proof is not limited to developing the content of a concept, provided it marks a real progress and attains a radically new object, then the dynamism of the intelligence which elaborates the proof implies finality. The mind is then 'commensurate' with the object in question. It is specified by that object beforehand. There is nothing accidental in the link between them. That is to say: by virtue of the something 'new' which it brings, an object of this kind is already present to the mind with a mysterious presence, a presence in germ as it were. Then, when it is grasped as the term of a logical process, when it is caught in the network of objective forms, it is, in a sense, 'recognised'. To demonstrate, in this instance, is 'to realise'. One 'discovers' what already was.

This is specially true of the proof of God. The finality essential to an intelligence which penetrates a new domain is then doubly unique. For in every other case, in fact, we are aiming at an object belonging to our own world, the world of experience, even if it is still beyond the grasp of our experience. But when, on the contrary, it concerns God, with reference to whom the very words 'object' and 'existence' assume a transcendental significance, it concerns the Being who is the source of my being, and who is 'more I than I myself'. How far above all others, and how much more intimate! In this instance, then, the procedure which accounts for the dynamism of the proof is a presence which has a stimulus and a profundity which are all its own, a sanctuary, the sign of God upon me, and that which makes me a spirit.[1] And at the same time, that which makes a person of me and makes me responsible. That is why the strongest of proofs depends more than any other—not of course in respect of its abstract form, but where its power of concrete persuasion is concerned—upon 'good-will'. For there is always more at work than the impersonal functioning of an intelligence. Charity and purity of vision are inseparable at this point from loyalty and honesty.

Furthermore, there is no essential heterogeneity between the spontaneous movement of the soul towards the recognition of God's existence and the rational analysis of the philosopher. Faced with the former, people are inclined to speak of instinct, heart, feeling, intuition: equivocal terms, all of them, which attempt to express the dynamism of the intelligence, its ultimate source, the unity of its movement, and at the same time to evoke the richly concrete and delicately sensitive region through which the light of the spirit makes its way. The philosopher's work is critical: he seeks to purify, to analyse, or sometimes to rectify or

---

1. Cf. St. Thomas, *Contra Gentiles*, III, c. liv, and Louis Lavelle's Preface to M. F. Sciacca's *L'existence de Dieu* (trans. Jolivet, 1951)

complete; but, above all, he analyses and decomposes that unbroken movement into its logical components and tries to check
and verify them. It is rare for him to pursue his studies beyond the
itinerary, to the heart of the dynamism, to that central and secret
point where reason and will originate. No doubt he is acutely
conscious that logic is no longer an adequate instrument of analysis; that one should go further, make suggestions, ask questions
and help the mind to a fuller awareness of itself; that one ought
to 'reveal', while always fearing to disturb, its latent content: a
delicate task which he regards as beyond his competence. And
then, perhaps, if it ceases to be a purely professional question,
so to say, if the problem touches him personally, it is possible he
will hesitate, fearing obscurely to meet not only a subject of
analysis, but God himself, not merely to discover the 'author of
nature', of the whole of nature,—but as a living man, to encounter
the living God, utterly unique and insistently at work in all men.
*Non enim fecit Deus et abiit. . . .*

                                        (V. Fontoynont, S.J.)

                                *

Where the proof of the existence of God is concerned, the
simplest classic form is, in itself, always the best.[2] It provides
the permanent scheme, which survives all the superficial
technical adjustments which each thinker, each age and every
school find it necessary to introduce. It continues feeding the
thought of those who think they can do without it—for 'the
proof which every man needs in order to attain full certainty
is so easy and so clear that one hardly notices the logical
process which it implies'.[3] That is what Fénelon calls 'a
sensible, popular philosophy open to any man free from passions and prejudices'.[4] In principle, as well as for the straight-
forward, honest mind, 'the merest glance reveals the hand

2. Cf. Régis Jolivet, *A la recherche de Dieu* (*Archives de philosophie*, vol. viii,
1931), p. 85.
3. Scheeben, *Dogmatique* (trans. Belet), vol. ii, p. 21.
4. Fénelon, *Traité de l'existence de Dieu*, I, i, No. 2.

that has made everything'.[5] Movement, contingence, exemplarity, causality, finality, moral obligation: the eternal categories are the starting-points open to man; they are always to hand and always resist his critique; they are as contemporary as man and his thought.[6] *Ecce coelum et terra: clamant quod facta sint.*[7] (Behold the sky and the earth: they cry out that they are made.) Or quite simply: *Aliquid est, ergo Deus est* (Something exists, therefore God exists.) 'The whole School is agreed that nothing further is necessary.'[8]

But if the spontaneous proof which springs up in this way is to impress reflective thought to the fullest possible extent, it will call for unceasing modification, and the resulting commentary will inevitably take the form of a justification, in some respects critical and never quite the same, by the nature of things. This 'learned' form of the proof, 'designed in the first instance to forestall and answer objections' implies a continual effort, constantly renewed, to adapt it to changing conditions. The need to adapt the proof will seem strange only to a man who has never dreamed of what is implied by the uniqueness of the case. 'The sublime and simple operation'[9] which leads to God remains fundamentally the same. The changes in technique, in perspective and presentation do not affect the proof itself.[10] God in his eternity dominates the

5. Fénelon, *loc. cit.*, No. 1.

6. The reader will have noted that we are here opposing the Kantian criticism and all that follows from it.

7. St. Augustine, *Confessions*, XI, iv, 6 (PL, xxxii, 811); *In Joannem*, tract. 106, n. 4 (PL, xxxv, 1910). Cf. *Wisdom*, 13: 1, 9.

8. André Bremond, S.J., *Une dialectique thomiste du retour à Dieu, loc. cit.*, p. 561.

9. A. Gratry, *De la connaissance de Dieu*, vol. i, pp. 45-46.

10. That is why it is possible to say that the proofs of God 'are not so much an invention as an inventory, not a revelation so much as an elucidation, a purification and a justification of the fundamental beliefs of humanity'; Maurice Blondel, *La Pensée*, i. p. 392.

incessant flux of creation, and in the same way the idea of God in us dominates the fluctuations of our intellectual life, imposing itself through those fluctuations with the same unalterable power. The great minds that have spoken about God are all our contemporaries.

\*

Kant tried to prove that the 'transcendent' use of causality was illegitimate, but the causality he had in mind was a narrow, scientific category, the specialised category of causation which rules the universe of Newton. Shaped for the ordering of phenomena, it exhausts its virtue in doing so. Kant's causality is, of course, only one example among many. In fact, modern Western philosophies 'are singular in one respect: the world they start out from is', as a general rule, 'the world constituted and constantly modified by the sciences'.[11] There is nothing surprising in the fact that this world is impotent, by itself, to provide a foundation for thought and sustain the movement of thought to the end. For that to be achieved, it would be necessary to dig down beneath the artificial, methodological categories of science to the great natural categories of reason. Then it might be possible to begin discussing the real question: on the one hand the negative critique, the view that the natural categories are illusory; and on the other hand a reflective effort to justify them and purify their spontaneous use.[12]

\*

Behind the apparent variations, the skeleton of the proof

11. Ferdinand Alquié, *La nostalgie de l'être* (1950), p. 151.
12. The book most helpful in freeing us technically from the Kantian criticism is probably Father Joseph Maréchal's *Le point de départ de la métaphysique*, cahier V (1926), p. 452: 'The transcendent principle of causality expresses this complementary and simultaneous revelation of objective contingence and of the perfection which measures it.'

always remains the same.[13] The proof is solid and eternal: as hard as steel. It is something more than one of reason's inventions: it is reason itself.

✻

All the objections brought against the various proofs of the existence of God are in vain; criticism can never invalidate them, for it can never get its teeth into the principle common to them all. On the contrary, that principle emerges more clearly as the elements with which the proofs are constructed are rearranged. That is because it is not a particular principle which the mind can either isolate and sift so as to determine its limits, or reject out of hand: it forms part of the substance of the mind. It is not a path which the mind can be discouraged from pursuing to the end, or one from which it can turn away, afraid of having taken the wrong road; path and mind are merged together. *The mind itself is a moving path.*

✻

*Causa essendi, ratio intelligendi, ordo vivendi* (the cause of being, the explanation of understanding, the pattern of living). All thought, like every being, and like every act, needs a principle and a term.[14] The mind did not set itself in motion, and its movement presupposes a direction; that is to say, a fixed point. The purely gratuitous is the purely absurd. One cannot do without the economy of God.

✻

13. Many writers have noted this without, however, explaining things in the same way. Cf. for example, Pedro Descoqs, S.J., *Praelectiones theologiae naturalis* (vol. i, p. 353; vol. ii, p. 15), and H. Paissac, O.P., *Preuves de Dieu, loc. cit.*, p. 88: 'The proofs of God radiate from a single centre: the affirmation of causality.'

14. St. Augustine, *De civitate Dei*, VIII, iv (PL, xli, 228–229); *Contra Faustum Manichaeum*, XX, vii (PL, xlii, 372).

If, as many people think, man's adoration of God were his adoration of humanity itself, he would adore it as nature or as an ideal; that is to say, as something realised or realisable. In either case, the object proposed would be no more worthy of adoration than the transcendent God such as he has been imagined to be and, so imagined, subjected to criticism.

If the divinity were conceived as a pure ideal, never capable of realisation, always becoming and never necessarily existing, by what right could it still be called 'humanity'? And in what sense could so elusive a term be called intelligible—or adorable?

These are three attempts to evade the living God, ways of escape into mystification.

\*

God is not the first link in the chain, the first of a series in the sequence of causes and effects which constitute the world.[15] God is not 'a point of origin in the past': he is 'a sufficient reason in the present' (in the past and in the future as well, and during the passage of time).[16] How many objections would disappear, how many misunderstandings would vanish, if that simple truth were understood!

\*

God is not merely the principle and the term, at the beginning and at the end: the Good of every good, the Life of all living things, the Being of all beings,[17] he is also at the heart

---

15. H. Paissac, *loc. cit.*, pp. 90–94: 'From the fact that there is causality in the world, it does not follow that there is causality for the world . . .' etc.

16. Etienne Gilson, answering Léon Brunschvicg in *La querelle de l'athéisme*, p. 228.

17. St. Augustine, *De Trinitate*, VIII, iii, n. 4: '*Bonum omnis boni*' (PL, xlii, 949). Pseudo-Dionysius, *Of The Divine Names*, i, 3 (PG, iii, 589), St. Bernard, *De consideratione*, book V.

of all things. *In illo vivimus, et movemur, et sumus.*[18] In him we live and move and have our being. But for that presence of the Absolute at the heart of the relative, of the Eternal at the heart of movement, everything would return to dust.

*

Becoming, by itself, has no meaning. It passes away and vanishes without really becoming at all: it is another word for the absurd. But without Transcendence, that is to say without a present Absolute installed at the heart of the reality which is in the process of becoming, not depending on it, but working within it, drawing it on, polarising it, making it really advance, there could only be unending becoming—unless a catastrophe were to come to put a violent end to everything, and the absurd were at last to rediscover its true nature, so to say, by becoming unequivocally nothing. . . .

All becoming is caused by Being. All becoming is turned towards Being. Becoming can only be thought by Being.

The idea of Progress, which magnifies and in some sort hypostatises Becoming, is one of the emptiest ideas which men have ever forged. Progress deified, it has been truly said,[19] is not only 'a race without a rudder', but a race without an end; or rather a race that gets lost without really being run at all. If you do away with the winning-post, you do away with the direction. The result is 'to create an abstract "beyond" that shimmers before the eyes of a distraught individual, a

18. St. Paul, in *Acts* 17: 28. Cf. John Scotus Erigena, *De divisione naturae*, i, n. 11; iii, n. 1 (PL, cxxii, 451–452, 621D). M. Blondel, *L'Action*, p. 346: 'God is at the centre of all that I think and that I do. . . . Going from myself to myself, I am continually passing through him.'

19. G. van der Leeuw, *L'homme et la civilisation*, in *Eranos-Jahrbuch*, vol. xvi (1948), p. 170.

will-o'-the-wisp that flies away at his approach'.[20] It is tanta-
mount to doing away with progress. 'To do away with absolute
perfection is to do away with any idea of becoming perfect.'
There can be no real improvement where there is 'neither
axis nor goal'; no real progression except by 'reaching the
limit'. If there is becoming, if progress is possible, then one
day there must be attainment (or let us say achievement); and
if there can be achievement, then there always has been something other than mere becoming.[21]

'Abolish the end of the world (which is also its beginning)
and there is no longer any *meaning* in things, only Chaos which
makes one despair and terrifies one, and to which Tathâgata
preferred *Nothingness*.'[22]

＊

On the one hand there is the absurdity of primordial chaos,
the nothingness from which everything is supposed to emerge,
which is said to engender being, the blind power which is
supposed to bring forth the light of the Spirit: and on the
other hand there is the source of Being—a certain 'Point
Alpha'.

On the one hand there is the hopelessness and the final
chaos of the ultimate defeat, of the Spirit finally overcome by
the darkness of matter, unending death, or that mournful
'eternal recurrence' in which all dreams finally vanish; on

20. Gaston Fessard, *France, prends garde de perdre ton âme* (1946), p. 149;
see pp. 133–150.
21. Félix Ravaisson, *La philosophie française au xixe siècle* (4th ed. 1895),
p. 50.
22. Paul Claudel, *Correspondance avec Jacques Rivière*, p. 60; published in
England under the title *Letters to a Doubter*. Cf. Plotinus, *Enneads*, v. i. 6:
'Everything that moves, requires something towards which it moves.'

the other there is the Place where being recollects itself—a 'Point Omega'.

'I am the Alpha and the Omega', says the Lord.[23]

✳

The intelligence, according to the philosophers of antiquity, 'is in some sense everything'. It is, indeed, spontaneously conscious of the fact, and whenever it attempts to articulate its dream, using the language of various systems, however strange and varied the formulae in which the dream takes shape, the intelligence is always concerned to understand everything in itself. *Vult autem anima totum mundum describi in se* [24] (The soul wants the whole world gathered into itself).

In other words, the intelligence cannot give up the Absolute for which it was made; but since it cannot situate or understand the Absolute, its natural reaction is to look for it in Nature, in the object that lies immediately to hand. But that kind of search must surely prevent the intelligence from attaining its end? The objective world is indefinite: an ocean without shores, where the mind is soon lost. To set sail there in the hope of someday dropping anchor 'beyond physical things' is surely to abandon the real world for the realm of abstractions? True metaphysics is the science *par excellence* of the real and the concrete.[25]

And so, at first, people believe in the data of the senses:

23. Cf. *Isaiah* 41:4; *Apoc.* 1:8. Cf. P. Teilhard de Chardin, *Le groupe zoologique humain* (1956), pp. 156, 162: 'The universal centre of psychical interiorisation'; 'An absolutely final principle of irreversibility and personalisation.'

24. St. Bonaventure.

25. Cf. Ferdinand Alquié, *La nostalgie de l'être*, p. 17: 'It is natural enough that most scientists, devoting their lives to the search for objectivity, should allow their need for being to be alienated in the course of their research; but the realism they profess is then part of a professional distortion.'

which have, after all, the privilege of being immediate. They brook no denial and survive no matter what theory. Are we not driven, in the end, to return to them?—Yet sooner or later one begins to notice that they are only appearances, or at the most the crust of reality. Then we place our trust in the entities fashioned by science; they, at least, provide these amorphous and fluid 'sensibilia' with a solid shell. They, surely, impose law and order.—But in the long run even the claims of science must be reduced. On closer inspection the entities which one took to be absolute appear contradictory, or can be resolved into yet others, such as movement, for example, or the 'atom' of antiquity . . . .[26] The scientific universe does not stand up to criticism any better than the sensual universe, unless it is supported by a universe of a different nature. The more successfully science and improved techniques bring the world under human control, the more does being, which cannot be brought under that control, evade us. . . . And in face of that new and apparently final defeat, the great temptation is agnosticism.—But agnosticism, which was conceived as a way of saving logic at least from the wreck, proves in its turn contradictory. The position is untenable. For how can one continue to affirm an Absolute which is admittedly unknowable? It seems, in fact, impossible to avoid complete scepticism.—But the intelligence can never abdicate. It cannot renounce its own formal law, it cannot cease judging, and that always means affirmation. Scepticism oppresses and undermines the mind because it introduces contradiction not only into the various contents of its various affirmations, but into the intelligence itself, into the heart of its every act. In trying to escape from that dilemma, the mind is sometimes driven to conceive (in the broadest sense of the word) a sort of *ersatz* Absolute Law. That introduces an inter-

26. *A-tome*: the uncuttable, that which cannot be disintegrated.

mediary sphere between the mind and the real half-way, as it were, between the immanent and the transcendent, a '*terra media* in which all our actions occur, and beyond which the need to know is lost in metaphysics; that is to say, in idle discussion and empty chatter concerning questions which have no possible bearing on practical life'. — But once again even that modest refuge proves unstable. Once again it has to be recognised that — just as the Absolute of the senses proved contradictory — the Absolute of Law is left hanging in the air. The 'eternal Axiom', by whatever name we call it, unless it is something else disguised, is ultimately the void, an abstract void without depth or mystery.

Here, surely, we have reached an *impasse*?

The original illusion is the cause of all the trouble. It arises from the uncriticised assumption that all we have to do is to perfect our knowledge of the world based upon certain primary data without bothering to reflect upon ourselves; from the blind assumption that the mind's vision is an extension of the body's vision, somehow prolonging its sight almost indefinitely, even when it seems, with the help of science, to sift the data and discover being beneath appearances; from the belief that the object, confusedly identified with being, must be amassed like a treasure, studied with a view to its usefulness and safeguarded so as to be enjoyed; in short, from the illusion that all we need to do is settle down in this world and become part of it. . . .

It is an illusion which is natural to the mind, as it is natural to man. Perhaps it is necessary; in any case, it is useful in encouraging the search for knowledge which is part of man's vocation. It is an illusion, nevertheless, which any thinking man will find the means to destroy in himself. He does so in a double way, by discovering that a perfectly adequate knowledge of this world is doubly impossible for him. For whether one calls it 'knowledge' or 'intuition'—according as one leans to a psychological explanation or a rationalist one —whether one conceives of it as a mysterious movement at the heart of the real as its forms dissolve or, on the contrary, as the living term of a great rationalist synthesis, as something immediate or as a construction—the ideal which seemed to activate human knowledge is a mirage.

Absolute Knowledge and the Intuition of the world are equally impossible.

Absolute Knowledge is impossible because its realisation would automatically involve the disappearance of the person who is to have it. He could not become the Knowledge and the Knower. All contradictions would have been overcome and oppositions cancelled out. All laws would fit together, one into the other, until finally they were contained in a single formula. And by that very fact all particular views would have vanished, the individual would have been dissolved in the universal, multiplicity would have been reduced to unity, and the formula would no longer discover a symbol in which to express itself, nor a consciousness in which to be affirmed. On reaching the end of his Knowledge, the Knower would be 'like the witch who ended by devouring her own inwards'. 'Nothing would remain but the unthinkable equality of nothing to nothing.'[27]

An Intuition of the world is no less impossible because it would dissolve the world which it is trying to embrace. There is an infra-intellectual element in this world, but if it could be fully assimilated and exhausted by the intelligence, it would no longer be itself. The ultimate reason for this must again be sought in the subjective realm: for if the world is essentially the world of sense, that is because it is essentially indefinite; and if the world is indefinite, and therefore inexhaustible and

27. Kierkegaard, *Journal*, XII, A, 354 (1850).

incapable of being reduced to a single total, that is surely because it is the necessary correlative of minds which are themselves in process of becoming.

In brief, the world is neither Law nor Essence. The antinomies which it always produces, stimulating and awakening the movement of the mind, will never be all resolved. Although real on their own level, the laws and essences which the intelligence goes on discovering will never be perfectly and absolutely clear. Science can never attain to that total synthesis which would make it identical with metaphysics, whose true and final object is not of this world. The understanding will never cease to be understanding; that is to say, an imperfect intelligence inseparable from the senses; but it is itself only a provisional substitute, an auxiliary of the spirit.

The understanding, the faculty of knowing, of science, is turned towards the outside; the mind, the spirit, must be turned inwards; there must be a 'recurrent critique' of thought, 'conversion', 'introversion', 'reflection', through which metaphysics at last discovers its own sphere. 'No,' Malebranche protested, 'I shall not lead you into a strange country, but I shall perhaps teach you that you are strangers in your own country.'

The understanding is open to an infinity of objects; a sign, surely, that it is open to the infinite itself. Without being able to sum them all up,[28] we can represent things to ourselves indefinitely; does that not mean that we desire, as far as is in us, to possess God? 'Acosmism' if you like; but in fact an acosmism which saves the world. Without it the world would be only 'systematic illusion'; thanks to it, the world is given back its value, its density, its meaning and its justification. It is revealed as a means, a stage—a trial. Its essentially indefinite character no longer scandalises us, and it can, in its present

28. St. Thomas, *Prima*, q. lxxix, a. 2.

guise, even slip through our hands, so to say, and vanish without disconcerting us—a transfiguration, the approach and proclamation of a better world. We are '*πάντα πῶς*', it has been said, because we are '*θεὸς πῶς*'; and it has also been said that we possess a 'faculty for the divine'; perhaps it would be more precise to say: *intelligence* is the *faculty* of being, because *spirit* is the *capacity* for God.

The human mind may be compared to a plant. The aim of the plant, in assimilating the elements which it draws from outside, is to live, to become itself. The aim of the spirit which first of all becomes understanding in order to assimilate the sensible, is not to lose itself in the elements which offer themselves to it, nor to use them to construct a self-contained and perfect edifice of knowledge; its aim is to become itself, to live. Its life is the possession of itself—and of all things—in that dependence upon God which illuminates it.

'I am among men,' says the traveller, 'not among angels, and I have no desire but for what breathes in my own image.'

'That is not true,' answers the voice, 'All your desire is for God, since the knowledge of God is your portion, and as the bee distils honey through the summer months, so your function is to contemplate the imperishable with loving eyes.'[29]

> *Noli foras ire, in teipsum redi, in interiore homine habitat Veritas; et si tuam naturam mutabilem inveneris, transcende et teipsum. Sed memento, cum te transcendis, ratiocinantem animam te transcendere. Illuc ergo tende, unde ipsum lumen rationis accenditur.*[30]

(Do not go abroad, return into yourself. Let truth dwell in the

29. Ernest Psichari, *Le Voyage du Centurion*.

30. St. Augustine, *De vera religione*, c. xxxix, n. 72 (PL, xxxiv, 154). Cardinal Zigliara, who quotes and comments on this text, is concerned to show that the doctrine of St. Thomas is in conformity on this point with that of St. Augustine. *Œuvres philosophiques* (French trans.), vol. ii, pp. 206–208.

inner man; and if you find your own nature mutable, then transcend yourself. But remember, when you transcend yourself, to transcend yourself as a ratiocinative soul. Direct your gaze, therefore, to where the reason's light itself is kindled.)

✲

*Solus Deus est, in quem nec pondus nec mensura cadit omnino, nec numerus. Unus Deus est, non habeat sui generis cui valet comparari.*[31]

(Only God is beyond weight and measure and number. God is one, and there are no members of his class with whom he can be compared.)

Everything that concerns God, everything that leads to God, everything that unites to God, is unique.

All 'communion' in name or in essence between God and other beings is excluded.[32] Nothing about God or our relations with him enter 'into a genus'.[33] There can be no exception to that principle. The way to God, however varied in its secondary forms and however numerous the ways become, is in itself unique[34] ἅπαξ.

✲

Polemon, Do you believe in one God?

Certainly, he answered; I believe in one eternal, self-existing principle.

31. St. Bernard, *De diversis*, sermo lxxxvi, n. 2 (PL, clxxxiii, 703B).

32. Marius Victorinus, *Adversus Arium*, iv, 23 (PL, viii, 1129D); St. Anselm, *Monologion* cc. 26, 27 (PL, clviii, 180A, B).

33. St. Thomas, *Prima*, q. iii, a. 5, etc. In *La philosophie de saint Bonaventure* (2nd ed., 1943, p. 115, n.), M. Gilson comments on the article quoted as follows: 'If God were in a genus something would be anterior to him; the idea of the genus is anterior, for the understanding which is classifying ideas, to that of the species contained under the genus' (English trans., p. 510). Cf. Cajetan, *In Primam*, q. xxxix, a. 1, n. 7.

34. J. Defever, S.J., *La preuve transcendante de Dieu*, in *Revue philosophique de Louvain*, 1953, p. 527: 'The proof of the existence of God is not the same as other proofs.' Cf. also J. M. Le Blond, *Le chrétien devant l'athéisme actuel* (*Études*, 1954, p. 301).

Whereas I, Callista replied, feel that God within my heart. I feel myself in his presence! He says to me, 'Do this: don't do that.' You may tell me that this dictate is a mere law of my nature, as to joy or grieve. I cannot understand this. No, it is the echo of a person speaking to me. Nothing shall persuade me that it does not ultimately proceed from a person external to me. It carries with it its proof of its divine origin. My nature feels towards it as towards a person. When I obey it, I feel a satisfaction; when I disobey, a sadness—just like that which I feel in pleasing or offending some revered friend. So you see, Polemon, I believe in what is more than a mere 'something'. I believe in what is more real to me than sun, moon, stars, and the fair earth, and the voice of friends. You will say, 'Who is he? Has he ever told you anything about himself?' Alas! no!— the more's the pity! But I will not give up what I have because I have not more. An echo implies a voice; a voice a speaker. That speaker I love and I fear.[35]

*

Thought can never reach being, though it fringes it with its very first steps. It would not move if it had not, in a certain sense, arrived.

*

The apparatus of the proofs is surely nothing but a vast *removens prohibens* — a clearing away of obstacles; all too necessary, indeed, in the carnal condition of the mind. But although its action is essentially positive when envisaged in the framework of the intellectual life, its significance is, above all, negative if it is replaced in the larger framework or the greater depth of the perspective of the spirit in its concrete actuality:

35. J. H. Newman, *Callista*, ch. xxvii.

The sculptor does not make the statue.
He removes what hid it.[36]

\*

The ways which reason adopts on its journey to God are
proofs, and these proofs, in turn, are ways. That does not
deprive them of their character as proofs—incomplete though
they may often be as proofs[37]; but their object, being unique
among all the objects of thought, confers a special character
upon them. They do not yield up their object, as other proofs
do, more or less. They do not enable us to penetrate it. God
alone is present to those who prove him, present in an intimate
manner—as he is to those who deny him. But at the same time
that presence is so beyond our grasp that he alone, among all
objects, cannot be held.[38]

\*

All men know God 'naturally', but they do not always
recognise him. A thousand obstacles, some inward, some external, hinder that recognition. Not everyone knows that he
knows God, and consequently not everyone does know him
'simply'. Thus, when I see Peter coming towards me—the

36. John Donne. Cf. Pseudo-Dionysius, *De mystica theologia*, ii (PG, iii,
1025). The idea comes originally from Plato, *Republic*, x, 611, and Plotinus,
*Enneads*, I, vi, 9.
37. Cf. the historical conclusions of F. van Steenberghen on the five ways
of the *Summa Theologica*: 'None of the *quinque viae* constitutes, as it stands, a
complete and satisfactory proof of the existence of God. The 1st and the
2nd need to be prolonged; the 3rd and the 5th need to be corrected and
completed; the 4th is unusable' (*Revuephilosophi que de Louvain*, vol. xlv,
1947, p. 168). These conclusions may, however, be discussed. Cf. William
Bryar, *Saint Thomas and the Existence of God* (Chicago, 1951).
38. Cf. Jacques Maritain, *Les degrés du savoir* (1932), pp. 445–456 (*The
Degrees of Knowledge*, p. 277); and St. Thomas, *Contra Gentiles*, IV, *i prooemium*.

comparison is St. Thomas's[39]—it is certainly Peter whom I
see in that being coming towards me, but I do not yet know
that it is he.[40]

It may be tempting to reject the comparison on the grounds
that it presupposes that I already knew Peter, and that I am
soon going to recognise the person I knew. Whereas in the
case of God, a 'proper' knowledge is to succeed to a knowledge which is still implicit. So I am not preparing to recognise him really, but to know him for the first time.

Of course no comparison can be perfect, and it is abun-
dantly clear that St. Thomas's does not apply in every particular, although it would be a mistake to minimise its bearing
too much. When I reach an explicit knowledge of God, I
certainly do not recognise him as someone whom I had
already known with the same sort of knowledge, and had since
forgotten or lost to view. I did not know him consciously, as
yet, in the ordinary sense of the word. Nevertheless, the
extraordinary thing is that knowing God for the first time I do,
in fact, *recognise him*.[41] For—to take up the illustration which
St. Thomas gives at this very point—when I come to know
God as the one who will make me happy, I realise at the same
time that God is identified with the beatitude which I knew

39. St. Thomas, *Prima*, q. ii, a. i, ad 1m.
40. I had written: 'When I see Peter coming towards me, without yet
knowing that it is he, as St. Thomas says'. Someone has observed that all
St. Thomas says is, 'I do not know Peter in the man coming towards me,
although it is Peter.' I do not see an abyss of difference between the two
translations. But I do take into account the words which immediately
precede: 'simpliciter cognoscere. . . '. In any case, I had no intention of
putting more into the French than the very simple but instructive sense
conveyed by the Latin of St. Thomas.
41. It is a similar paradox that makes St. Augustine exclaim, in words
addressed to Reason: 'Qui nondum Deum nosti, unde nosti nihil te nosse
Deo simile?' (*Soliloquies*, c. ii, n. 7; PL, xxxii, 873).

by desiring it, but which I placed, at first, among objects which deceived me; or rather I can now identify my beatitude with him. That is certainly recognition. And that is what always happens. I never discover the existence of God as I might discover some distant city, for example, to which I was not bound by any real tie and which I should only note as an external fact. That is what Father Jules Lebreton means when he says that 'to speak of God to a human being is not to speak of colours to the blind.'[42] Many men, no doubt, behave with regard to God like the blind in respect of colours, but the problems of reflective philosophy must not be confused with those of psychology or sociology; and when the blind man recovers his sight, then the moment he knows God it would be true to say that he has recognised him. For—and this is the extraordinary and really admirable thing—'the habit of God', belonging as it does 'to the very nature of the spirit', is possessed by it 'before any act whatsoever'. That is what St. Thomas implies by his comparison, and when he explains, at the same time, that we do not know God 'simpliciter' with our first knowledge, which is purely 'natural' and implicit, his qualification means that in a certain sense God is, nevertheless, known; that is why, when the moment comes, it is permissible to speak of 'recognising' him.[43]

\*

To believe in *an eternity in the instant*—as all spiritual experience of a high order demands—without admitting that it is a participation in *eternal Eternity* is to plunge into contradic-

42. *La connaissance de foi* in *Études*, vol. cxvii (1908), p. 735.
43. Cf. H. Paissac, quoted above, c. ii, note 29. It is clear, moreover, that the text of St. Thomas quoted is not by itself enough to authorise the *whole* doctrine concerning the implicit knowledge of God.

tion. It involves living on an illusion without fully admitting it.

\*

One can, of course, bring the classic objection of the hundred thalers against St. Anselm's argument: existence is not a predicate, it is not a perfection of the essence. . . . One might also object that the being to which his argument leads is ultimately nothing but the thought which posits it; it would only be valid in so far as it contains the 'virtuality of idealism'.[44]

For, 'how can that infinite, which can only be thought of as objective, exist if not as the mind's own power of transcendence? What can that unsurpassable greatness be if not the greatness of thought? And by that very fact, the thing affirmed, which does not require that thought should leave its own realm to make the affirmation, seems only to be thought affirming itself by affirming its superiority to everything. There we find being absorbed into thought.' So that by concluding to the existence of God distinct from that thought, one would be realising a particular case, and a very typical one, of the 'alienation' which the heirs of Hegel are so persistent in denouncing on all sides.

That objection reveals the full significance and all the implications of the argument better than anything else could possibly do. All that is needed is that the Anselmian dialectic should be carried to the point at which it rejoins the classic 'argument from contingence'.

44. D. Parodi, *Le rationalisme et l'idée de Dieu* in *Revue de métaphysique et de morale*, 1930, pp. 41–42: 'In the only form in which it can still be admitted by the critical metaphysics of our time, the transcendence of the Absolute can only be the creative energy of thought, realising in some way or other the consciousness of its own unity and continuity, of its fecundity and its infinite progress.'

From a strictly and exclusively intellectual point of view, the argument achieves and affirms nothing except the thought which affirms itself, and which alone can discover itself to be unsurpassable through its power of effacing all images, and destroying all representations, and of surpassing all limits. Then reflection, which becomes conscious of its unlimited greatness, is inevitably faced by an ironic ambivalence: is it other than I? Is it possible from that strictly noetic point of view to settle the question? We believe so, because the intellectual experience, which alone is entitled to intervene at this point, is not only an experience of the triumph of reflection by which thought affirms itself superior to all else, but also an experience of the labour of reflection, of the act of pushing back and denying the limit, which always needs to be renewed. The experience is not only one of greatness, but also one of littleness . . .; greatness and littleness are given together in their relationship which is the act itself of reflection and of transcending. But because the limit is crossed, because the unattainable is affirmed, it is impossible to forget that the transcending requires the limit and is also consciousness of the limit. From that I know that I only am through the unattainable, and at the same time that I am not the unattainable. I suffer from the greatness which is my condition and which makes me recognise my limitation. What is affirmed as being, both in the individual mind and in reality at the same time, is certainly absolute thought; but that absolute thought, which the finite spirit cannot appropriate or absorb entirely, and which the mind must therefore oppose to itself, becomes, in that opposition, the reality of thought, that is to say being itself. . . .[45]

From a strictly intellectual point of view, then, it cannot be said that the Anselmian proof contains the germ of idealism and immanentism; it is a meditation on the power and limita-

45. Jacques Paliard, *Prière et dialectique, méditation sur le 'Proslogion' de saint Anselme*, in *Dieu Vivant*, 6, pp. 56 ff. And the same author's *Sur un aspect de la structure scientielle* in *Actes du 3e congrès des sociétés de philosophie de langue française* (Bruxelles, 1947).

tions, on the wretchedness and greatness of our thought taken together. It does not lead to or justify an illegitimate 'aliena- tion' of man; it shows him, by its recognition of his limitations, the secret of the only way of surmounting them.

*

The proof of God is incomparably stronger than any other proof, because, more than any other, it is bound up with the mind which propounds it; but it is also—as experience teaches—easier to elude, and these apparently contrary characteristics spring from the same cause. For God is not an object among other objects. If he is, he can only be the total Object and the total Truth informing the whole mind. Now in rejecting *a* particular truth, one only accepts *an* absurdity; whereas in rejecting the total Truth, one intro- duces absurdity itself into the mind. As long as the intelli- gence adheres to some solid region of being, the least absurdity naturally horrifies it, since the least absurdity once recognised is enough to unsettle the inner coherence of its mental uni- verse; whereas once the law of contrast ceases to operate, the mind finds great difficulty in 'realising' total absurdity, which is presented to it in the form of a sort of inverted coherence, co-extensive with its whole knowledge. Total absurdity infects that knowledge fundamentally but without disturbing its internal relationships. Thus the intelligence can always delude itself with its own subtleties.

*

The proofs of the existence of God are continuously sub- jected to two sorts of criticism the details of which may some- times coincide, though they are as different in origin as in their results.

The first of these criticisms is inspired by a strict but narrow conception of the intelligence; and if they are directed upon the proofs formulated in the past, they envisage their conceptual apparatus and their 'logical forms' in a superficial way, without regard to the spirit that informs them. Criticisms of this kind are a form of historicism which becomes so literal and precise that the essentials of the doctrines which they set out to interpret escape its grasp; it attributes a certain poverty of thought to the proof which is nothing but the poverty of its own method. . . . Criticisms of this type always lead to doubts, on the rational plane at least, about God's existence, and at the same time mutilate the intelligence itself.

The second type of criticism, on the other hand, springs from the exigencies of faith in God. It will have nothing to do with proofs which do not lead to the true God. It does not want a Cause, an End or a Legislator whose transcendence is not assured. It is guided by the same superior instinct which motivated the proofs. It is not really a critique so much as a deepening of the proofs. It conspires with them and helps them to rectify and perfect themselves. It uncovers their real nature. It discerns and distinguishes their motivating principle. Through them the mind becomes conscious of all that is unique and complete in the various proofs, and of what gives them a force above all other proofs. Its achievement is to bring the proofs to the point at which they become one; although their expression may be marked by a particular mentality or by the state of the sciences at a particular time, the result is to bring out the supereminent and eternal validity of the activity of the mind which, without seeing God, infallibly posits his existence. That activity involves no constraint upon the mind; on the contrary, the mind cannot refuse it except by doing violence to itself and, in so far as that is possible, destroying itself.[46]

*

46. Cf. M. D. Chenu, *op. cit.*, p. 153.

The professional who wonders 'whether the proof of the existence of God can be popularised' would seem to hold that only a few specialists, technicians in the 'science of metaphysics' are within their rights in affirming the existence of God with a full understanding of the issues. They alone, it would seem, are in possession of true certainty. Everybody else would be under an illusion in the matter. Intellectually, their affirmation would not be valid. At the most they might be said to benefit in the more or less remote 'preparations' for the proof itself, 'very useful, provided they are taken for what they are worth'.

How is it that those who argue on these lines do not see that if they really believe in God they are claiming an exorbitant privilege? Or at least that they are only right from a superficial point of view?

There is, M. Jacques Maritain tells them, a 'doubly natural' knowledge of God which is the fruit of the apperception of being, 'much profounder than any logical process scientifically developed' because it has its roots 'in a single and primordial intuition'.[47] Knowledge of this kind does not make the 'scientific' proofs superfluous, but on the contrary makes them possible, since that is the basic testimony which supports the proofs to which, ultimately, we must always return.[48]

❆

There is often entertainment to be had from concentrating on certain details of our 'learned' proofs, so-called. The

47. *Approches de Dieu* (1953), pp. 10, 18; cf. pp. 15–16.
48. Cf. L. B. Geiger, *Bulletin de Philosophie* in *Revue des sciences philosophiques et théologiques*, 1954, p. 268: Recent works in natural theology 'mark, at bottom, the abandonment of the Wolfian type of rationalism, in which metaphysical concepts must be separated as far as possible from all empirical data, and consequently also from the whole pre-philosophical life

materials used are, indeed, not uniformly solid[49]; the categories on which they rest may not always have been sufficiently
well tested; the dialectical apparatus in which they are set out
may be obsolete, and the subtle objection may slip through its
meshes. 'Sometimes [critical reflection] finds itself in the
presence of a rich and profound thought which is still only
implicit and unexpressed; at other times, on the contrary, it
comes up against explicit formulae which claim to be authentic proofs of the existence of God, and then the formulae
almost always seem vulnerable at some point or other.'[50] In
short, there is no guarantee that the believer who reasons will
necessarily be rigorously logical, a competent analyst, an
up-to-date scholar or a profound philosopher. Even if he
reasons well, his technique may be poor. There is no reason
to be ashamed of acknowledging the fact.

The point requires further consideration. Scholarly thought
is in fact technical thought, and as such it is artificial in the
etymological sense of the word. Now artifice, though legitimate, can always be countered, provisionally at least, by
another artifice, even if a sophistical one. In order to answer
the objection, greater precision and technical justifications
will become necessary, while their full value may only emerge
of the mind. On the contrary, it seems to us important to emphasise the
whole spontaneous movement by which man rises up to God. The truly
philosophical ways have no need to fear recalling their humble origins.'

49. Cf. F. van Steenberghen in *Histoire générale de l'Église* (ed. Martin and
Fliche), vol. xiii (1951), p. 254, n. 7, where the author takes an example
from St. Thomas. Also M. Chossat, S.J., art. *Dieu* in the *Dictionnaire de
théologie catholique*, vol. iv, col. 932–935: 'As for the argument from the
prime mover as St. Thomas understood it, it has long since ceased to be
taught, even by Thomists. . . . [It has only] survived in Protestant scholasti-
cism, among some well-meaning philosophers and apologists.' (These reflections might be dubbed 'historicist'!)

50. F. van Steenberghen, *Revue philosophique de Louvain*, vol. xlv (1947),
p. 166.

in the light of new points of view. And as a result new objections will certainly be brought forward. As Fénelon says, after making oneself 'understood by the ignorant', it still remains to 'answer the rash criticism of men who misuse their minds against the truth'[51]—but also to recognise the justice of new demands which may yield a positive gain. After the difficulties propounded by Locke and Hume, for example, came the difficulties, engendered by them and yet so different from theirs, put forward by Kant. Following upon these came the difficulties put forward by Hegel and so many others after him, always to some extent unforeseen—and if it is to be adequate, the answer presupposes reflections of a kind which were not previously current. And so on without end. Reason is never at the end of its tether. The dialectical chain always forges new links for itself. Mind on the march is never secure against a false step, and goes astray into many an *impasse*, but at the same time it digs down deeper within itself, and invariably discovers new sources: its life never ceases. It is an illusion to imagine it can ever be satisfied with itself.[52] And the only way of not becoming stagnant is to make the best of it. It is never possible simply to rest on the achievements of the Ancients, not even on their happiest efforts, not even if one makes every effort to assimilate their work thoroughly.

51. *Lettres sur divers sujets de métaphysique et de religion*, Letter ii.
52. How can anyone have thought that it was perhaps 'reserved for our century to arrive at fully satisfactory formulations' of the proof of God? 'That, it has been said, has nothing surprising about it if it is admitted that metaphysical knowledge in our age has for the first time reached the end of the process of its historical genesis.' But that, in spite of Hegel, is precisely what one cannot easily admit. Péguy with his sound sense said: 'One does not go beyond Plato.' Nor, we might add, beyond St. Thomas. And on the other hand, if we are not to lose the essential, which is always threatened by new methods, new efforts must be made: the illusion noted here appears to us to add the illusion of the historical to the illusion of the definite.

That does not mean that we should disdain them or persuade ourselves into thinking that we have outstripped them; far from it. But mere repetition is not the best way to recover their thought. The proof, in any case, remains fundamentally the same: one never gets beyond it. But the primary certainty has always to be recaptured, and in order to re-establish the simplest truths one must be prepared, in certain cases, not only for a long struggle but for unforeseen discoveries.

But in the meanwhile those who really believe in God will not allow themselves to be troubled. No learned objection will be able to shake their faith—whether the objection comes from the rational order, the dialectical, the psychological or what you will. The reason for this is that the artifice developed by the learned proof is, for them, simply the elaboration and the rational organisation of a permanently subsisting proof, at once simpler and more fundamental, a proof which is natural, spontaneous and in many cases unformulated but nevertheless inscribed 'in the deepest recesses of our reasonable nature', a proof which never ceases, even when the objections seem unanswerable, to engender a perfectly reasonable conviction 'stronger and more unshakeable than any conviction artificially formed',[53] a proof, in fine, which is the indestructible mainspring of learned demonstration.

So, in the matter of God, whatever certain people may be tempted to think, it is never the proof which is lacking. What is lacking is taste for God.[54] The most distressing diagnosis

53. Scheeben, *Dogmatique* (trans. Belet), vol. ii, p. 21. Cf. *supra*, notes 3, 8. Cf. Maritain, *Approches de Dieu* (1953), p. 16: 'The knowledge of God, before being developed in logical and perfectly conceptualised demonstrations, is first and foremost a natural fruit of the intuition of existence' (English trans. 1954, p. 8).
54. Cf. H. Geursten, *Les preuves de l'existence de Dieu* in *Dixième congrès international de philosophie* (Amsterdam, 1948), vol. i, p. 838: 'The value

that can be made of the present age, and the most alarming, is that to all appearances at least it has lost the taste for God. Man prefers himself to God. And so he deflects the movement which leads to God; or since he is unable to alter its direction, he persists in interpreting it falsely. He imagines he has liquidated the proofs. He concentrates on the critique of the proofs and never gets beyond them. He turns away from that which convinces him. If the taste returned, we may be sure that the proofs would soon be restored in everybody's eyes, and would seem—what they really are if one considers the kernel of them—clearer than day.

✳

'Every country has its sources of water, but the Philistine with his earthly tastes did not know how to find water everywhere. He did not know how to find reason and the Image of God in each human soul.'[55]

of the argument does not depend upon our voluntary acceptance of it, but we consider that the inclination to accept it is the essential condition for perceiving its intellectual force.'
55. Origen, *On Genesis*, Homily xiii, n. 3 (PG, xii, 232–233).

At the conclusion of these two chapters devoted to the affirmation and the proof of God, in the course of which we have, at several important points, drawn on the doctrine of Joseph Maréchal and his disciples, we think it may be useful to quote two passages from his essential work, *Le point de départ de la métaphysique*, cahier v. These two passages sum up and provide a foundation for what precedes, and will help to explain part of what remains to be said in the following chapter, 'The knowledge of God'.

The perspective opened on the internal finality of the intelligence as a foundation for our analogical knowledge.

. . . But the objection springs up again as we dispose of it: all our concepts, according to the Thomist doctrine, are originally concepts of material quiddities: the contingence of created being—as revealing the divine transcendence—is given to us neither in the representation which these concepts contain, nor in their abstract and universal form: the former is only a diversified relation to the phantasm, the latter a process of objectified generalisation, which does not go beyond the level of being to which the representation belongs. There is no trace, it would appear, of a 'transcendental relation' in the objective concept, such as would be metaphysical contingence of finite being.

One must admit the objection if the concept is only representation and mere generalising abstraction. How, in fact, could the absolute and transcendent term of the contingent relationship, God, reveal itself, even by means of analogy, in finite representations or in the mere generalisation of them? Does not St. Thomas tell us that the culmination of our knowledge of God is 'to know that he exceeds everything that we could conceive about him'? The transcendent God cannot therefore be represented by our concepts, nor even guessed at as the limit towards which the generalisation of our concepts trends. We should only become conscious—it would appear—of the radical contingence of created objects by escaping their finitude, by becoming conscious of the absolute super-eminence of their principle over all other possible objects of our thought. But what might that consciousness be if it is neither an intellectual intuition, nor an analysis of an intuition of the senses, nor the abstract consideration of a material form—and on what could it rest?

Thus we are faced with postulating, in our objective knowledge, something more than the static reception and the abstractive analysis of 'data'; we are driven to postulate a movement of thought continually carrying us 'beyond' what can be represented by concepts; to postulate a sort of metempirical anticipation which would show us the objective capacity of our intelligence to expand infinitely until it surpassed all limitation of being. Apart from that there can be no analogical knowledge of the transcendent. To explain and safeguard it we are therefore led to place ourselves on the path of the *dynamic finality* of our minds; for it is only the intelligence's 'internal finality' that can enable it continually to surpass the present object and indefinitely pursue a greater object. . . . (Pp. 184–185.)

The reader will not have failed to note a certain resemblance between this interpretation of the transcendent principle of causality—which we believe to be traditional—and the metalogical procedure of Jacobi and Père Gratry. According to these, the apperception of a finite object involves the affirmation of the Infinite, latent in the depths of our intelligent nature in a virtual state. That affirmation need not necessarily be perceived, but would exist in the very texture of our objective thought, or at any rate a careful analysis could reveal it.

Is the doctrine of Jacobi, or at any rate that of Père Gratry, true in essentials? Or is it false? We prefer to say that it is incomplete and ambiguous.

Something at the heart of our particular apperceptions must determine or necessitate the affirmation of the Absolute; otherwise we shall never be able to demonstrate analytically, starting from finite objects, the existence of the Infinite. But here, it would now seem, is the point at which a distinction becomes essential: the affirmation of the Infinite either is or is not—as implicit affirmation—a *constitutive* condition of our apperception of particular objects.

*If it is not*, if it is only a concomitant or a subsequent condition in relation to the primary object of our understanding, the finite object, it would imply only an act of 'rational faith' in the transcendent absolute on the occasion of each of our apperceptions and by virtue of a subjective necessity; the transition to the infinite would be a real 'a priori synthesis', natural, legitimate even, but not justified by a genuinely speculative or scientific necessity; the understanding might, in fact, avoid the transition without becoming involved in logical contradictions.

*But if it is*, if, that is, the affirmation of the Infinite is a logical condition which preceded and really constitutes the apperception of finite objects, then, certainly, we can neither deny it nor withdraw it without flagrant contradiction. But it would be necessary to add, under pain of slipping into ontological intuitionism, that the transcendent affirmation which is the dynamic constitutive condition of the object thought of, has nothing in common with a 'vision of objects in God' nor with an 'innate idea' even a 'virtual' one in the Cartesian sense. Purely implicit and 'exercised' in the apperception of finite objects, it can be made explicit only dialectically, by reflection and analysis.

Let us try to guess what Jacobi or Gratry meant and perhaps incorrectly expressed. All that we need to conclude from the alternative given above is the following: if we opt for the first answer, in other words, if we leave the implicit affirmation of the divine Absolute outside the intelligible structure of the finite object, we shall be opposing Kant's agnostic conclusions with the simple fact of our instinctive impulse of the reason, an impulse —how ironical!—which Kant is the first to proclaim; on the contrary, if we opt for the second term, then we shall have embarked in practice on an absolute and radical refutation of Kantian agnosticism, starting from the methodological exigencies of the *Critique* itself. (Pp. 452–453.)

# 4

## THE KNOWLEDGE OF GOD

THE universe through which God reveals himself is not only his work: it is his creature. It is not merely a thing which God in his omnipotence made out of nothing; and for that very reason it is a being which exists and lives only on the life and the being which it is continuously borrowing from its Author. Or rather—since the classical metaphors of 'loan' and 'source' are either too feeble or too strong—the universe lives and exists only in God. *In Eo vivimus et sumus.*

God is 'his own being', but he is also 'the being of all'.[1] He is incomprehensible, inaccessible and at the same time familiar and close to us. 'The root and principle of every creature', he is the Being present *par excellence*.[2]

God comes to us on all sides through the world; it is his Being that comes to solicit our attention. We ought to be able to meet him anywhere and recognise him everywhere. Whether we consider the 'great world' or the 'little world', the cosmos that surrounds us or our own spirit, everything real that comes within our orbit is by the whole of itself, and first of all by virtue of its existence, the symbol and the sign

1. Pseudo-Dionysius, *Celestial Hierarchy*, iv, 1 (PG, iii, 177–178). John Scotus, *De divisione naturae*, i, n. 3 (PL, cxxii, 443B); n. 72 (518A). St. Bernard, *De consideratione*, V, vi, n. 13 (PL, clxxxii, 796A).
2. St. Thomas, *Prima*, q. viii, a. 1. John of St. Thomas, *Cursus Theol. In Primam*, q. xliii, dissert. xvii, a. 3–4–11. St. Bernard, *In Cantica*, sermo iv, n. 4 (PL, clxxxiii, 798A–B).

of God[3]; not an artificial sign of some kind or other, deliberately chosen and valid by convention, but a natural and, for us, a necessary symbol. It is an ontological sign which we cannot discard or emancipate ourselves from. God is not seen directly or apart from a sign; but God can be seen everywhere, through the world, however obscurely. Every creature is, in itself, a theophany. Everywhere we find traces, imprints, vestiges, enigmas; and the rays of the divinity pierce through everywhere.[4] Everything is drenched in that unique Presence. Everything becomes transparent 'to a pure gaze and a steady eye'.[5] If our knowledge, like our ignorance, troubles our contemplation, if our mind's eye cannot see beyond the outer shell of the world, if we are open to nothing sacred—or if, on the contrary, the world seems 'full of gods'—that is because our sight is blurred. It is only too true; the world conceals God more than it reveals him. Things have become opaque. And yet it remains true that the Creator 'has scattered the reflections of his divine perfections upon his creatures, and that thanks to those visible lights we are able to know, by analogy, the splendours of the inaccessible Creator'.[6] *Invisibilia Dei per ea quae facta sunt intellecta conspiciuntur*[7] (The invisible

3. Cf. Godefroy de Saint Victor, *Microcosmus*, xl (ed. Philippe Delahaye, 1951, p. 61).

4. Cf. Pius IX, *Lettre à Mgr. de la Bouillerie* (Castel-Gandolfo, 30 July, 1864) in *Symbolisme de la Nature*, by Mgr. de la Bouillerie.

5. Paul Claudel.

6. Th. de Régnon, *op. cit.*, vol. iii, p. 458.

7. St. Paul, *Rom.* 1 : 20. St. Augustine, *Confessions*, VII, xvii, n. 23; x, n. 16; IX, x, n. 24. Hugh of Saint Victor, *Didascalion*, VII, iv; St. John of the Cross, *Spiritual Canticle*, etc. Such texts indicate a long tradition of thought and exegesis. It would be, in fact, to limit the sense of the Pauline assertion if we were to exclude from it what may be called the symbolic or exemplarist element, seeing in it only an indication of a proof of existence. 'It was not until the Cartesian period that symbolism was ejected from theology and reserved for the use of mystics' (M. D. Chenu, *op. cit.*, pp. 48–49).

things of God are made known by the things which are made).

'O Thou who appearest through every form and structure without adhering to them or being confounded with them!'

From this it follows, in the first place, that the knowledge of God which comes to us through the external world is itself, in a sense, a revelation:

> The greatness and the beauty of creatures
> Make us, by analogy, contemplate their Author.[8]

It is an objective revelation just as natural reason is in itself, as we have seen, a subjective revelation. There is a double and unique natural revelation, a gift of the sign and of the capacity to interpret it, a gift of the book and of the capacity to read it. For it is not my mind which first rises from the world to God: it is God who first descends, in some sort, through the world to my mind. However spontaneously it may come to me, the proof comes only in the second instance. The proof is my construction, but the sign which precedes it and already contains it, which allows it, provokes it, sets it in motion and always overflows it, is given me by another. In all truth, God *makes me a sign*.[9] The first language he uses to communicate with me is his creation. Being created by the Word, everything which comes from him is a word and speaks of him. It is for me to attend and to answer—but the initiative is not mine.

By that very fact, such knowledge of God the creator, though

---

8. *Wisdom*, 13:5. Cf. Leo XIII in the encyclical *Aeterni Patris* (*Acta Leonis* XIII, Rome, 1881, p. 268). St. Thomas, *In Rom.*, c. i, lectio 6.

9. God manifests his wisdom through his visible creatures as by so many signs—just as one man communicates his thought to another by the signs of language: St. Thomas, *In Epist. ad Romanos*, c. i, lectio 6. St. Maximus Confessor (PG, xci, 1328). Cf. M. E. Boismard, O.P., *Le Prologue de saint Jean* (1953), pp. 111 ff: 'The first phase of divine revelation by the Word was the work of creation. . . .' Cf. H. Urs von Balthasar, *Phénoménologie de la vérité*, pp. 86–87.

always mediated, is not entirely indirect. To borrow a word from St. Augustine, one might call it *contuitio*. God, in a sense, invests me with his signs, and I perceive him in his creation— until such a time as I can see his creation in him. But that knowledge is always obscure[10] owing to the weakness of my intelligence,[11] though it is none the less concrete, in principle, even when it follows the paths of logic and abstraction, because it is the knowledge of a *Presence*. Reasoning by itself, supposing that the initiative were entirely mine, reasoning to which I was not provoked, and which was not the result of a stimulus or an essential impulse, would give me only a completely indirect and wholly abstract knowledge. It would supply me only with a pure concept, taking the place of an *absent* being—or rather an absent thing. Yet in fact, the true God reveals himself as present under cover of the abstraction which derives from me.

Indeed, if, *per impossibile*, God were present in me only as a pure abstraction, in the form of a concept, then granted that any representation of God is totally inadequate, it might be asked how I could possibly avoid agnosticism.[12] Several people have asked the question without finding the answer. On such a hypothesis, my knowledge would be empty. But in fact the hypothesis is empty. For *there is always more in the concept than the concept itself*,[13] and that is what the critics of the concept

10. St. Bonaventure, *In Hexaemeron*, collatio xiii, n. 14.

11. St. Thomas, *De Veritate*, q. v, a. ii, ad. 11m.

12. As can be seen, the phrase is not in the indicative, and does not refer to the real. It expresses a hypothesis, which it declares at the same time false and impossible. In the language of the schools, we should say it is a *videtur quod non*, which is bound up with the answer to it. One must apologise for giving such explanations, which are unnecessary for most readers.

13. Cf. Joseph Maréchal, *Nouvelle revue théologique*, 1931, pp. 193, 197. And in the same sense, André Bremond, *loc. cit.*, p. 575. And Etienne Gilson, *Le Thomisme*, 5th ed., p. 67 (English trans., p. 44).

forget, as well as many of its defenders. That is what enables
me to understand that the knowledge of God remains concealed beneath the need to criticise any representation of God.
All the negations which accumulate and appear to be purely
destructive are, in the end, at the service of an affirmation
both stronger and purer than they. If God conceals himself, it
is in his very presence. His transcendence does not mean that
he is exiled from the world; it is the exact opposite of an
absence.[14] Every creature reveals him by virtue of the being
it borrows from him, crying out that it is not he. Such is the
mystery which, in spite of its obscurity, is always a light; the
emptiness which it demands of us is the form of his Fullness.[15]

   *Patet quam ampla sit via illuminationis, et quomodo in omni re
quae sentitur sive quae cognoscitur, interius lateat ipse Deus*[16] (It is
clear how broad is the way of illumination, and how, in
everything that is perceived or known, God himself lies
hidden).

                                *

   The hidden God, the mysterious God, is not distant and
absent: he is always the God who is near.

                                *

   Unable to think of God as he is, we take the course, which is as
wise as it is humble, once it is properly understood, of conceiving
him as like ourselves. We take the qualities which we attribute
to him from our relation to him, and as those relations are real,

14. Cf. M. D. Chenu, *op. cit.*, p. 161.
15. Etienne Gilson, *op. cit.*, pp. 121 (English trans., p. 83), 147 (English
trans., p. 101).
16. St. Bonaventure, *De reductione artium ad theologiam*. St. Anselm, *Monologion*, c. xiii (PL, clviii, 161A–B). St. Augustine, *De Musica*, VI, c. xiii,
n. 40 (PL, xxxii, 1185).

what we say is true, although as definitions they are inadequate to the point of being worthless.[17]

It is a wise course, and the knowledge it gives us is true knowledge, because the course we follow is inspired by a deeper thought and because our knowledge is lit by a more secret light.

\*

Those who uphold immanence deny transcendence, whereas those who believe in transcendence do not deny immanence. Indeed, they grasp the idea of transcendence sufficiently to understand that it necessarily implies immanence. If God is transcendent, then nothing is opposed to him, nothing can limit him nor be compared with him: he is 'wholly other' and therefore penetrates the world absolutely. *Deus interior intimo meo et superior summo meo . . .*

The champion of exclusive immanence thus reveals his partiality. Only the champion of transcendence is impartial, like the truth itself.

The champion of immanence refuses to consider transcendence except as a spatial image dominating a thought which is thenceforward purely mythical. He really claims the exclusiveness of the *logos*. And yet unless he is prepared to end up with the absolute emptiness of being and thought, he remains the prisoner of his imagination, no less than his opponent: the 'inside' is no less spatial than the 'outside' or the 'beyond'. *A priori* there is no reason, for example, to regard it as 'the concave spatial aspect' of a sphere. . . . And if 'height' is an imaginative concept, so is 'depth'. And however subtle the form, immanentism itself might well be an 'inverted spatial

17. A. D. Sertillanges, O.P., on St. Thomas, *Somme Théol.*, *Dieu*, vol. iii, p. 343; cf. p. 340. Cf. St. John Damascene, *De fide orthodoxa*, I, iv (PG, xciv, 800).

prejudice'.[18] From whichever angle one considers the question, a critique of the imagination becomes necessary.[19]

*Locales quidem excedit (Deus) temporalesque angustias, sed libertate naturae, non enormitate substantiae!*[20] (God exceeds the bounds of time and space, but through the supreme freedom of his being, not through the size of his substance!).

The champions of immanence are nourished by the illusion that by 'interiorising' the faith of those who believe in transcendence, they become the ultimate interpreters of a prophet who did not see where his intuition was leading him. In that way they believe they can ensure the transition from belief to philosophy and from spontaneity to reflection. In their disdain, wedded to a comprehensive indulgence—of which the history of thought offers many an example—they imagine they can justify the believer and criticise him at the same time. They think they can transform a relative truth into an absolute truth. But the attempt, also witnessed by the history of thought, to find a new 'over-all' concept, a new 'beyond', a new transcendence within immanence itself, an attempt which is always being renewed and is always unsuccessful, bears

18. The expression was used by Gabriel Marcel in the debate already referred to (*La Querelle de l'athéisme*, published in *De la vraie et de la fausse conversion*, 1950): 'I cannot help fearing that this cult of interiority . . . is, in spite of everything, an inverted spatial prejudice.' Etienne Gilson had denounced the same 'tiresome confusion' in Léon Brunschvicg.

19. And this criticism will perhaps finally recognise that the elevation and the movement of transcendence are not so much the primary data of sensibility, subsequently transposed by a questionable analogy of an imaginative order, as values directly perceived, by means of a spontaneous ontological symbolism, in the object of sense itself. For the history and use of the words 'transcendent' and 'transcendental', see Jean Wahl, *Traité de métaphysique* (1953), pp. 642–649. Cf. St. Augustine, *De diversis quaestionibus*, lxxxiii, book 1, q. 29 (PL, xl, 18–19).

20. St. Bernard, *De consideratione*, V, xiii, n. 28 (PL, clxxxii, 805B); St. Augustine, *Confessions*, I, iii.

witness again in favour of the believer. Transcendence always has the last word. And the victory of transcendence consolidates immanence—the two ideas are "as though woven together"—though its significance is altered.

> God is wholly within and wholly without, supereminent and yet intimate, surrounding us yet present in us.
>
> There is nothing above him, nothing outside him, nothing without him. Beneath him, in him and with him are all things. . . .[21]
>
> He is everywhere in his entirety, one and the same . . . penetrating all things by engulfing them, engulfing them by penetrating them. . . .[22]

\*

> I went into the higher part of myself, and higher still I found the Kingdom of the Word. Impelled by curiosity to explore still further, I descended deep into myself, and yet I found him deeper still. I looked outside, and met him far beyond everything exterior to me. I looked within: he is more inward than I myself.
> —And I recognised the truth of what I had read, that we live and move and have our being in him. . . .[23]

\*

'The function of art', Léon Bloy wrote, 'is to fashion gods.' It is the function, too, of thought, and of the higher realm of human activity which is that of the 'poet'.

But in addition to the *poet* who fashions the gods, there is also the *prophet* who receives the revelation of God. The sculptor of idols who gives form to the gods is balanced by

21. St. Hilary, *De Trinitate*, I, vi (PL, x, 29), commenting on Isaiah 40:12 and 66:1. St. Augustine, *Soliloquies*, I, i, n. 4 (PL, xxxii, 871). Cf. St. Ambrose, *De Fide*, I, xvi, n. 106 (PL, xvi, 553).

22. St. Gregory the Great, *Moralia in Job*, II, xii, n. 20 (PL, lxxv, 565).

23. St. Bernard, *In Cantica*, sermo lxxiv (PL, clxxxiii, 1141B–C).

the iconoclast who refuses to allow God to be immured in a
form. There is the intellectual who organises his thoughts into
a body of thought, and there is the mystic who rejects them as
they take shape, or rather from whom they are gradually
withdrawn.

This opposition between different types of men is surely
one of the fundamental aspects of the conflict between Hellen-
ism and Judaism which can be followed down to modern
times. But it is also an opposition within man himself which
cannot be resolved, a tragic but fruitful confrontation. Here
it is impossible to make a definitive choice without sacrificing
something essential. The poet is a man who 'has dreams' and
who 'declares the visions of his heart', who is always in danger
of becoming a 'false prophet' if he proclaims them as truths
received from above.[24] Nevertheless, the prophet needs the
poet—and is a poet himself in his own way—because man
cannot receive anything into his mind without collaborating
with his own thought: even the object of revelation must,
after all, be *conceived*. In the same way the mystic needs the
intellectual because detachment from defined forms—*videntur
ut paleae*—presupposes the work which constructs those forms,
and the judgment which acknowledges their value. So the
conflict cannot end in the victory of either side. It must be
transformed into harmony. It must become a rhythm—and
mirror the rhythm first sounded by the incarnation, death and
resurrection of the Saviour.

✱

Things *mean* God; but they do not 'express' him; nothing can
express him but himself. Spirit, too, would express him and never
succeeds. But it affirms his existence as the presence of a limit
beyond our reach, towards which the world in the course of its

24. *Jer.*, 23: 16–28.

existence is for ever in movement, a sort of marvellous obstacle which cannot yet be seen, but is announced and indicated beforehand by a signal. The spirit alone has a presentiment of its end deep within itself, for it is made for and launched upon the path to God and can become conscious of its movement. Although that consciousness hardly counts until the things of the world make their signs.[25]

\*

The further one gets from the *He*, the more vividly one realises the *Thou*—and through the *Thou* its necessary correlative the *I*. Immanence and exteriority develop together. Reflection, far from dissolving or annihilating personal being, gives it a foundation. That is the principle misunderstood by idealism which has never conceived of any relationship except that of a knowing 'subject' and a known 'object'.

But the exteriority thus established is not, as can be seen, something 'objective': it exists within an 'inter-subjectivity'. It is not the exteriority of a *He*—which would soon be degraded into an *it*—but that of the *Thou par excellence*. It is not the exteriority of an object which one dominates—which can be annihilated in imagination—but that of a subject to which one gives oneself, in which one finds oneself, which one has to think of as subsistent. That subject is in truth the *Other*, in the strongest possible sense of the word: the absolute Other, the mysterious Being enclosed within itself, always beyond our grasp, the totally personal Being, 'the only *Thou* which, by definition, cannot become *That*'.[26] He who cannot be represented, but whose reality is all the more compelling; through the knowledge of whom we become conscious of ourselves, and through love of whom we possess ourselves; whereas, as long as one clung to the representation, it only

25. Hubert Paissac, O.P., *Le Dieu de Sartre* (1950), pp. 75–76.
26. Martin Buber, *I and Thou*.

bred an illusory 'other' without inwardness, without mystery and without real fruitfulness for the subject. Idealism was then right perhaps to reduce that 'other', by absorbing the object in the subject or the world into the representation—without succeeding, however, in providing a basis for an *I*, a real personal Subject.

There is no unique subject: no personality without otherness; no consciousness turned in upon itself; no real being without intersubjectivity[27]; no real knowledge nor ontological density without mystery. And no man without God.

*

Are we really to believe that if 'many Christians' like to see their God as the 'Thou' *par excellence*, it is because they 'no longer dare to say "He"', as they do of a being'? Is it true that 'by making God "Thou" in the dialogue we call prayer, they compound with the view that the structure of our consciousness requires the presence of others, and so only appears to attain God as part of the human'?[28] The criticism is too subtle and the analysis very imperfect. Why should the speaker to whom one says 'thou' be 'a part of the human' any more than the object addressed as 'it'? Why should dialogue find its explanation in ourselves any more than representation? If there is 'intentionality' in all knowledge, which modern philosophy stupidly ignored and which contemporary phenomenology is helping it to rediscover, it may well be that this 'intentionality' is expressed more forcibly in dialogue which puts the subject into direct relationship with another subject. In any case, prayer—whether it is petition, adoration or abandonment—is not just any sort of dialogue; the reciprocity which it implies does not imply equality, as the believer who

27. Gabriel Marcel, *The Mystery of Being* (English trans., Harvill Press).
28. Ferdinand Alquié, *Solitude de la raison*, in *Deucalion* I, p. 188.

turns to God in prayer knows well enough, even if he has no aptitude for philosophical reflection. In point of fact, those who speak of God—rightly or wrongly—as a 'Thou' rather than as a 'He' are not betraying their timidity. Even if they insist on this distinction in too exclusive a spirit, their purpose is certainly to safeguard the idea of the divine transcendence by refusing to speak of Being as 'a being'. When they establish the relationship of prayer with him, they affirm both their own dependence upon him and his independence of them. By refusing to immure him in the 'object' their intention is clearly to liberate him from what is human.[29]

\*

If God is already known in some way in our knowledge of duty, (even by those who think themselves unable to see him and call themselves atheists), it may be said that God is found and possessed in some way in the fulfilment of duty.

However, that can be said only upon one condition. For there are two ways of acknowledging duty and consequently two ways of fulfilling it. In order that the knowledge and possession of God which goes with the knowledge and fulfilment of duty should not remain purely implicit,[30] duty must be regarded not as a purely formal law in the Kantian sense, but as the *requirement of the Good*. Only then can there be any question of getting beyond that abstraction 'the natural law' and of making room for the real God, even in the case of those who do not yet know his name.

29. Cf. S. L. Frank, *God with us* (English trans. 1946), p. 79.
30. It should be noted that this does not apply to all natural knowledge of God, but solely to the knowledge in question, implied by knowledge of moral duty when it is fully grasped. The fact that there is knowledge of God both natural and explicit is stated in other contexts. Cf. Yves de Montcheuil, S.J., *Dieu et la vie morale* in *Mélanges théologiques* (ed. 1951), pp. 141–157.

'Independently of the realisation in action of any explicitly conscious knowledge', the intelligence has therefore 'a vital and non-conceptual knowledge (of God) enveloped in the practical notion—vaguely and intuitively grasped, but with all its intentional power—of the moral good as the formal motive of the first free act, and in the movement of the will towards that good, and at the same time towards the Good itself'. By committing himself and choosing, man 'thinks of what is good and of what is evil; but by the same token he knows God without being aware of it, because by virtue of the inward dynamism of his choice of the good for the sake of the Good, he desires and loves the transcendent Good as the final end of his own existence'.[31]

In doing so he transcends the idea of a 'natural' God, the author and extrinsic sanction of the Law, or mere pattern of moral value. For the Good, which *ex hypothesi* the moral agent recognises in practice as a demand upon himself, the Good to which he secretly gives his adherence, is, as we have just seen, the 'Good which subsists separately', and nothing else. It is the 'ultimate Good', in reality God, 'in whom, whether he knows it or not, he places his last End'. It is the God who *is* that last End, and therefore the God of grace.

In other words, the God who reveals himself implicitly in this way to the man of good will—still unable to name God— is the God who is not content to command man, but draws him and desires to draw him even to himself, the God, according to St. Thomas, for whom man makes his first free choice when he turns towards the good,[32] in such a way that 'the first deliberate act of the will, the first act of the moral life, in the

---

31. Jacques Maritain, *Raison et raisons*, pp. 137–139 (English trans. *The Range of Reason*, pp. 69–70).

32. Cf. St. Thomas, *Prima Secundae*, q. lxxxix, a. 6; *De Veritate*, q. xxiv, a. 12, ad. 2m.; *De Malo*, q. v, a. 2, ad 8m., etc.

strict sense of the word, is steeped in the mystery of grace'.[33]
That is the God who, in Pascal's words, 'makes the soul feel
that he is its sole good and that it cannot rest except in the
love of him'.

✳

The greater the mystery, the more the imagination will see
its irreplaceable role grow, once it is recognised for what it is.

We know, for example, that our God is a personal God. For
'being has the countenance of a person.'[34] There is no temptation to turn back to the nature myths of so many religions,
nor to their philosophical counterparts. But once this is
admitted and clearly established, it remains true that analogies
from 'nature' are more evocative than 'personal' analogies—
simply because they are more distant. They may even help to
purify the analogical method. The analogy from nature is 'the
natural cure and the normal complement which our feeble
spiritual development requires'.[35]

In order to understand the importance of symbol and meta-
phor where knowledge is concerned, one must bear in mind
the law that suggestion is in inverse proportion to the definition, and the value of an evocation is in inverse proportion to
the value of thought.[36] One must also remember that law in

33. Cf. Jacques Maritain, *Neuf leçons sur les notions premières de la philosophie
morale* (collect. 'Cours et documents', 1951), pp. 123, 127.

34. André Marc, S. J., *Dialectique de l'affirmation* (1952), p. 605. Cf. Louis
Lavelle, *De l'Acte*, p. 63: the person expresses 'the deepest essence of
being'.

35. Charles de Moré-Pontgibaud, S.J., *Sur l'analogie des noms divins, l'analogie métaphorique*, in 'Recherches de science religieuse', 1952, p. 182;
cf. pp. 166, 173.

36. Cf. Frithjof Schuon, *De l'unité transcendante des religions* (1948), p. 77
(English trans., p. 81). Certain consequences which the author draws from
this principle belong to his special system.

order to appreciate the language of the mystics. Then we shall not be disturbed by a progression which, to our logic, seems to move in reverse. 'O Father, Spouse, Brother', the mystic exclaims; then, as the movement of his contemplative love reaches its term, 'O deep, calm River, devouring Fire, Light of all lights!'[37]

\*

To be honest, one would be hard put to it to say precisely which of the 'divine attributes' was intended by the word 'light'. In the first instance, no doubt, it evokes the limpid transparency of a knowledge which penetrates the whole of reality without effort —but it also qualifies the purity of an essence which the shadow of nothingness has never darkened: 'God is light and there is no darkness in him.' But in addition to the qualities which may be called 'external', the word 'light' helps us to realise the expansive power and the radiation of a sovereign Spirit, in the absence of which all other spirits would be benighted, and in a more general way to realise the triumphant, superabounding manifestation which we call 'Glory'. 'Light' communicates fluidity and splendour to all these aspects of infinite Being which conceptual analysis and definitions threaten with extinction; it binds them together, develops them and enfolds them in the mystery of beatitude.[38]

\*

O God, you are Love indeed—but Love which is different from my love! You are Justice indeed—but Justice which is different from my justice! If I am lacking in love and wanting in justice I shall inevitably stray from you, and the worship I offer you will be neither more nor less than idolatry. To believe in you I must believe in Love and in Justice, and it is a thousand times better to believe in them than simply to call upon your name.[39] Apart from them, I can never hope to find

37. Pierre Lyonnet, S.J., *Écrits spirituels* (1951).
38. Ch. de Moré-Pontgibaud, *loc. cit.*, p. 167.
39. Cf. St. Bonaventure, *In Sent.*, d. 8, q. I, q. II, concl.

you, and those who take them as their guides are on the road that leads to you. But to adore you in spirit and·in truth, and to avoid the risk of adoring myself, I must, furthermore, believe that my justice—even that justice which I conceive of without being able to realise it—is not yet your Justice, and that my love is not yet your Love. My ideal is not your reality. When I apply the words Justice and Love to you, I still do not understand you.[40] For 'we know imperfectly and prophesy imperfectly', and everything remains an enigma to us—we can make an idol of Justice itself, and perhaps even of Love itself.

For God is above all names[41] and all thought, beyond every ideal and beyond all value![42] A Living God!

✼

The affirmation of God rises up from the very roots of being and thought, before all conscious acts, before the formation of concepts, conferring upon consciousness its guarantee and upon the concept its universal validity. Shrouded and secret, though necessary and permanent, it lies at the basis of all our judgments about being. It is one with the life of thinking being. It is the affirmation of God which gives the thinking being, at all levels, coherence and consistency—without which it would vanish into dust, just as without God the world itself would vanish. *Forma mea, Deus meus.*[43]

But in order to rise to consciousness and become, on its own level, a judgment among other judgments, that fundamental affirmation must objectify itself. It does so in a thousand

40. St. Thomas, *Prima*, q. xiii, a. 5.
41. St. Augustine, *Contra Adimantum Manichaei discipulum*, xi (PL, xlii, 142).
42. St. Augustine, *Sermo* 341, c. vii, n. 9 (PL, xxxix, 1498).
43. Cf. St. Thomas, *Prima*, q. cvi, a. i, ad 3m.

imaginative forms, and finally expresses itself in the definite form of a concept, the necessary instrument of human thought. But that necessary instrument, even where used correctly, is also necessarily inadequate. The Absolute, upon which our knowledge rests, now enters into the system of our knowledge. It therefore appears to be caught up in the universal network of relationships. And the Transcendent, which by definition 'goes beyond' the notions elaborated by our intelligence, seems to allow itself to be imprisoned within them.

At that point the equally necessary labour of intellectual purification begins—instinctively at first and then methodically. For how, in fact, are we to save the idea of God unless we can save the idea of the Absolute and the Transcendent? As the idea of God is particularised and becomes objective, it is submitted to a negative dialectic, which is turned against all the gross elements from which it seems to take its substance.

But is there no danger that, in the end, the process of purification might empty the affirmation of meaning? Might not that be the price of not falling into idolatry? For in the last analysis 'it is not as a concept that God would have us think of him, nor even as a being whose content would be that of a concept.' As the object takes form, Being evaporates. . . . 'Beyond all sensible images and all conceptual determinations God affirms himself as the absolute act of being in its pure actuality.'[44] Any element that can be grasped, whatever it may be and however fine the filter through which it has been passed, will always be too gross to express the Being whose essence is Being, pure Being, pure 'Existence', pure Act undefiled and simple, a Subject which cannot be a predicate,

44. E. Gilson, *L'esprit de la philosophie médiévale*, p. 52 (English trans., pp. 52–53).

which must be posited 'absolutely', and in respect of whom a 'qualification' is a sacrilegious 'restriction'.[45] How, in these circumstances, can we say anything about God? How can we know anything about God? Any likeness which we think we can note between Creator and creature is at once annulled and more than annulled by a far greater unlikeness.[46] Are we not obliged to confess that 'what God is remains totally unknown to us'?[47] So that when we have at last eliminated all risk of idolatry, thanks to a thorough purification, shall we not fall into agnosticism—can we avoid the feeling of becoming atheists?

As we have seen, that is mere appearance. The intelligence cannot retrace its path, and unless it lies to itself, it cannot go back upon a necessary affirmation. Although by a paradox, the effect of which we cannot help feeling, what prevents us reaching such a conclusion is precisely what produces the appearance of it. What, we may ask, is a pure affirmation? What is an affirmation which, to all appearances, affirms nothing, since it does not proceed from a subject distinct from itself, and does not bear upon any object? 'He Is': how can that Fullness, if reached by analysis alone, not seem at first sight empty?

The proof, once again, is not for that reason invalidated. The demands of reason remain unchanged. The affirmation continues to be the centre of thought, tenuous perhaps, but tenacious.[48] It continues to rise up again and again, always necessary, always imperious, always invincible, and despite

45. A. Bremond, *Une dialectique thomiste* . . . *loc. cit.* pp. 572–573, and cf. Gilson, *L'Être et l'Essence* (1948), p. 326.
46. Fourth Council of the Lateran (1215): 'Inter Creatorem et creaturam non potest tanta similitudo notari, quin inter eos major sit dissimilitudo notanda.'
47. St. Thomas, *De Veritate*, q. ii, a. i, ad 9m.
48. St. Augustine, *De libero arbitrio*, II, xv, n. 39 (PL, xxxii, 1262).

all the criticism and the scruples which seek to prevent its
emergence, always prepared in the long run to engender the
same process of objectivisation. Being fortunately incapable
of fusing within their limits any of the particular forms in
which it tends to take shape, it suffices to prevent the intelligence from ever settling in the negations which follow upon
them—which inevitably follow.[49] And by the same token it
closes the road to retreat. The affirmation allows neither of
denial nor of doubt, and cannot be called in question. There
is no appeal from the judgment of reason. What is gained is a
permanent acquisition.

It may seem, perhaps, as though man were destined to
oscillate for ever between those two poles without ever finding a haven in which to rest. Reason may remain serene with
its proofs untouched; but the man who reasons is perplexed.
The theoretical problem may well have been solved in principle, and all appearances to the contrary may have been overcome in theory; but the practical question remains, the fundamental question about the use of the idea of God in the spiritual
life.

Then, to our astonishment, the Gift of God intervenes. It
is the second gift, for the first is the gift of mind itself, of the
affirmation. Then the *donum perfectum* is grafted on to the *datum
optimum*. It in no way adds to the force of the affirmation or to
the value of its rational foundation. It in no way supplements
the proof, nor does it offer a substitute, which is not, in any
case, required. Its action is of a different order: it comes to
assure the spirit of the tranquil enjoyment of its object, without depriving it of its original impulse, and to restore peace
to the mind. And precisely because it is 'of another order,

49. It should be noted that the oscillation is not between the affirmation
and the negation of God, between dogmatism and scepticism, but between
the two moments of knowledge: affirmative theology and negative theology.

supernatural', a situation which had seemed inextricable is unravelled without effort. It is as though a new dimension had been introduced. By allowing us to participate in the Life of God himself, the life of charity furnishes our idea of God with a spiritual content. Instead of interrupting or discouraging the process of purification which, on the contrary, it may even stimulate, this spiritual content helps us to pursue the work in peace; because, on another plane, it ensures at a single stroke the peaceful continuation of that work and, if one may so express it, confers an indispensable density upon the affirmation.[50]

That Gift is the Spirit of God—in fact the 'Spirit of Jesus' —through whom our hearts are filled with charity. 'No one, strictly speaking, knows God unless it be God alone. The Spirit enables us to know him in a certain measure, because it assimilates us to him. . . . The Spirit alone can plumb the depths of God. The Spirit alone grants us a knowledge of God which is more than radically inadequate or purely negative knowledge. The Spirit makes new men of us, men participating in the divine nature, as the Second Epistle of St. Peter makes bold to say, and gives us the only knowledge of God which is on his own level, because it is a connatural knowledge.'[51]

But this does not mean that the natural laws of the intelligence are suspended. The increase in knowledge obtained in this manner is not of a rational or philosophical order. It is

50. This does not, of course, mean that the affirmation owes its intrinsic solidity to grace. The value of the affirmation in itself, in an impersonal state, so to say, is one thing; the serenity or repose of a particular mind in which the affirmation triumphs is another thing. Cf. the last works of Maurice Blondel, or Karl Adam, *Christ our Brother*, and St. Thomas, *Prima*, q. xxxii, a. 1, ad 3m.

51. Louis Bouyer, *Le sens de la vie monastique* (1950), pp. 132–133 (English trans., *The Meaning of the Monastic Life*, p. 83).

something more or something else: but however that may be, it is different. It is not the privilege of the scholar or the reasoner. It is a knowledge related to experience.[52] Or rather, it is worthless apart from that experience—which is wholly spiritual and beyond the reach of psychology. In one respect it is the privilege of that personal intimacy and concrete intuition which belongs to all religious knowledge, but in another respect it participates in its extra-scientific character. It is a simple, quasi-immediate knowledge, although in reality it is always analogical, *in speculo*.[53] For 'he who loves', St. John says, 'is born of God, and knows God'.[54] 'He who loves', Augustine comments, 'sees love, and he who sees love sees God': *inde videmus, unde similes sumus* (there we see where we are like); and this love, adds William of Saint-Thierry, is the eye which enables us to see God: *ipsa [caritas] enim est oculus quo videtur Deus*.[55] But at the same time, that knowledge is precarious and always obscure, for it depends upon a life which is both precarious and full of 'vicissitudes', and never possessed as one possesses a natural good; and because it can never catch the pure light which is shed by that life, in the prism of the concept—and indeed never attempts to do so.

*Putas quid est Deus? Putas qualis est Deus? Quidquid finxeris non est; quidquid cogitatione comprehenderis, non est. Sed ut aliquid gustu accipias, Deus caritas est, Caritas est qua diligimus*[56] (Do you think what God is? Do you think what God is like? Whatever you imagine, he is not that; whatever you can grasp in

52. Cf. St. Thomas, *In primum Sent.*, d. 14, q. ii, a. 2, ad 3m.; d. 15, q. ii, ad 5m.

53. Cf. St. Thomas, *Contra Gentiles*, III, xlvii, and St. Augustine, *Confessions*, XII, c. xxv, etc.

54. *In Jo.*, 4:7.

55. William of St. Thierry, *De natura et dignitate amoris*, cap. vi, n. 15 (PL, clxxxiv, 390B).

56. St. Augustine, *De Trinitate*, VIII, viii, n. 12 (PL, xlii, 957–958).

thought, he is not that. But to give you some taste of him, God is charity, and Charity is that by which we love).

*Novimus haec (de Deo). Num ideo et arbitramur nos comprehen-disse? Non ea disputatio comprehendit, sed sanctitas: si quo modo tamen comprehendi potest quod incomprehensibile est. . . .*[57] (This we know (about God). Do we therefore suppose that we have understood? It is not argument which makes us understand these things, but sanctity; if indeed what is incomprehensible can be in any way understood. . . .)

Such, indicated schematically, are the principal stages in the dialectic of the idea of God, of its *concrete* dialectic in the life of the *concrete* spirit. Always supported by the first, the other four stages are engendered, follow one another, merge, conflict and are harmonised in the ever-moving complexity of this idea which is stronger than all criticism—stronger than death.[58]

\*

Our power of affirmation is greater than our power of conception or our power of argumentation. For although the last two may be called in question at some point or other, the first remains intact, and revives the others. It also enables them to reach a successful conclusion.

Hence an ebb and flow and a whole series of apparently irre-concilable conflicts—which nevertheless, provisionally at least, always find a solution. What we can least prevent ourselves from affirming almost always seems compromised in the realm of argument and concepts. The critical reason is endlessly inventive in discovering arguments against what had been most solidly established. Among those who make use of it, some are mainly concerned with our idea of God while

57. St. Bernard, *De consideratione*, V, xiv, n. 30 (PL, clxxxii, 805D).
58. Cf. Father de Vries in *Scholastik*, 1950, i, p. 129, where the author develops the same idea.

others attack our proofs. From a purely logical point of view it is no easy matter to show that they are wrong, since the believer is often awkward when he gives a rational justification for his belief; his philosophy may be short-winded, his analysis inadequate; moreover, his belief may oblige him at times to accept the criticisms with both hands. Yet nothing prevails against our affirmation which always restores the value of our ideas and our proofs. Our affirmation is ceaselessly inventive against the inventiveness of criticism.

For God is *the Wholly Other* and in every sense. A process capable of leading us to other beings and to other truths could not, of itself, and as such, lead us to him, any more than the representations which fittingly express other beings and other truths can, by themselves, express him. His mystery remains inviolate even when logic has obliged us to affirm that he exists. Our reason does not penetrate into him.[59] Dialectic and representation cannot cross the threshold. But reaching out beyond dialectic and representation, our spirit affirms him who, although reached by their mediation, is beyond their grasp. And that affirmation, as it passes from darkness into light, and then again from light into another darkness, always remains invisible.

As created spirits, we are impelled towards the Absolute. Many things conceal us from ourselves and tend to deflect that impulse. But deep within us it persists, waiting to be released. So that when we apply ourselves to the business of criticism, correcting the process of our thought and its products, we are being faithful to our nature and to the impulse which makes us what we are. Criticisms can neither deflect nor obstruct it: for the impulse itself inspires them, directs them and gives them a positive meaning. And it is in that impulse that the Absolute makes itself known to us.

✳

59. St. Thomas, *In librum de Causis*, vi.

The philosopher and the spiritually-minded, the primitive and the civilised man, the most personal thinker and the humblest of believers, the 'prophet' and the mystic, do not merely converge in their use of the word: God. If they are correctly orientated they really do meet, however partial and sometimes narrow their outlook may be, or at least they tend to meet—and in that tendency do meet—although the object about which each of them is thinking seems to be different.

They have one idea of God—and of the soul—in spite of its many and diverse origins, in spite of the differently formed concepts, and in spite of representations which differ strangely.

It is as though they shared a single space, one external world, although within that world there are distinct worlds of sight and smell, sound and touch. . . .

God is indeed unique! And the astonishing convergence of so many points of view, each one of them seemingly independent, is a further testimony to his uniqueness.

*

God of the intelligence and God of the consciousness— God of supernatural revelation and God of reason—God of nature and God of history—God of being and God of value— God of reflection and God of prayer—God of the philosopher and God of the mystic—God of the soul and God of the universe—God of social tradition and God of solitary meditation . . . so many opposites and one unity!

Infinite God and perfect God—perfect in his infinity and infinite in his perfection! God who is both Absolute and personal!

God who is unique under so many aspects; the one End approached by innumerable ways. God of my whole self! God of all men! None of the avenues which lead to you is closed, and I have no right to forbid the use of any one of them.

*Voces diversae, semitae multae: sed unum per eas significatur, unus quaeritur*[60] (There are many voices and many paths, but the same thing is indicated by them and the same person sought).

\*

I do not believe we need think of God as a bearded old man, and I think we can form a different idea of him. It is in that direction that the sacred frontier of the spirit lies, where man, leaving his senses behind him, like Moses removing his sandals before the burning bush, and like Jesus leaving his three disciples behind him, goes to pray a little farther off, 'at a stone's throw', armed solely with his heart and his intelligence. That is where metaphysical awe begins, the 'ecstatic aphasia' spoken of by Plotinus. And in spite of the fact that we cannot express this idea, how much more intense, how much more vivid it is than the idea we form of an ordinary object![61]

\*

The idea of the Good, the Prime Mover, Necessary Being, the One superior to being, the Universal Principle, Nameless and Formless Deity; God of the Patriarchs, God of Moses and Isaiah, Sovereign Lord, Judge, King of History, Father of Jesus. . . . Between each of them an abyss—and yet they are, or at least can be, the same God.

\*

'There are many outside who appear to be inside, and many inside who appear to be outside.'

The words used by Origen and Augustine[62] are apt at all

60. St. Bernard, *De consideratione*, V, xiii, n. 27–29 (PL, clxxxii, 804–805).
61. Paul Claudel to Jacques Rivière, 11th May, 1908. *Correspondance de Paul Claudel et de Jacques Rivière*, pp. 158–159.
62. Cf. *Histoire et Esprit*, *L'intelligence de l'Ecriture d'après Origène* (coll. 'Théologie', 1950), pp. 158–159. St. Augustine, *De Baptismo contra Donatistas*, V, c. xxvii, n. 38 (PL, xliii, 196).

times. The fact that they can be misused should not be
allowed to conceal their truth. And what is true of belonging
to the Church is, doubtless, no less true of belief in God.
One can be an atheist and profess belief in God[63]; one can be
a believer and call oneself an atheist.[64] *Novit Deus qui sunt ejus*
(God knows who are his).

✳

'Those who scrutinise the Majesty will be overawed by the
Glory.'[65]

The philosopher does well to look with suspicion upon
titanic metaphysical structures. He should not imagine that
he can rise, of his own power, to a genuine 'science of God'.
He should use his critical faculties to moderate the pride of
his curiosity. If, at the conclusion of his efforts, he begins to
rediscover the object towards which his deepest impulse
moved him, and affirms the existence of God, he is simply
giving a principle of unity to all beings, a basis to his thought,
an explanation of his own existence and meaning to the
universe. Thus he limits himself 'to putting down the only
answer required by the world in question: God has not yet
unveiled himself'.[66] And though he may pursue his reflections
beyond the proof, they will never enable him to penetrate
into the divine nature. His inkling of the unknown may perhaps
keep him outside it. Nor will all the wisdom of which he is

63. 'There is an atheism concealed in all hearts, which is diffused in all our
actions: God counts for nothing' (Bossuet, *Pensées détachées*, ii).
64. This does not mean that all atheists are unconscious believers! Cf.
Maurice Blondel, *La Pensée*, vol. i (1934), pp. 392–393.
65. *Prov.* 25: 27.
66. H. Paissac, O.P., *Théologie, science de Dieu*, in *Lumière et Vie*, i (1951),
p. 36.

capable allow him to begin contemplating God himself; he can only contemplate 'the economy of his wisdom'.[67]

\*

'Lord, I do not try to reach your heights, for I do not put my intelligence on your level. But I long for a glimpse of the truth which my heart loves and believes in.'[68]

67. Evagrius, *Gnosticos*, V, 51.
68. St. Anselm, *Proslogion*, c. i (PL, clviii, 227B).

# 5

## THE INEFFABLE GOD

INFINITE intelligibility—such is God. The incomprehensible is the opposite of the unintelligible. The deeper we enter into the infinite, the better we understand that we can never hold it in our hands. *Quidquid scientia comprehenditur, scientis comprehensione finitur*[1] (Whatever is understood by science is limited by the understanding of the knower). The infinite is not a sum of finite elements, and what we understand of it is not a fragment torn from what remains to be understood. The intelligence does not do away with the mystery, nor does it even begin to understand it; it in no way diminishes it, it does not 'bite' on it: it enters deeper and deeper into it and discovers it more and more as a mystery.

At the summit of man's effort, face to face with the Being of God, the nothingness of man is brought home to him; and in the same way, as the Mystery of God allows itself to be penetrated by reason—or rather as it penetrates reason—it reveals its depths, and the light which it radiates only increases the obscurity in which the Mystery conceals itself. That does not, strictly speaking, mean that we realise increasingly 'the infinite distance between God and man'—to use Kierkegaard's expression—as though God withdrew his greatness from us in proportion as the infinite grows in us, and as we come the better to see that the divine is not 'simply the super-

1. St. Augustine, *De civitate Dei*, XII, xviii (PL, xli, 368).

lative of the human'.[2] It is not really a question of distance, or at least the word only expresses one aspect of the reality. God does not retreat to the outermost edge of our horizon; he does not escape us entirely, or better still he does not let us escape him—but in this, too, he reveals himself as God, as the incommensurable, the inapprehensible, that is to say impregnable. One can, therefore, without the least fear, enjoin reason to 'understand'. But with each advance and each time it discovers some new wonder in the divine Object, the desire to know and the desire to comprehend will be mortified.[3]

The exercise of 'speculation', says the Carmelite, Dominic of St. Albert,[4] 'is the deepest death that a loving spirit can suffer.'[5] And the Angelic Pilgrim, Angelus Silesius, adds:

> The better you know God, the more you agree
> That you are less and less capable of expressing what he is.

*Amictus lumine sicut vestimento.*

(Clad in light as in a garment.)

✳

The mind which tries to 'comprehend' God is not a *miser* amassing a heap of gold—a summa of truths—which goes on increasing. Nor can it be compared to an *artist* returning to a rough sketch, adding to it, improving upon it and, in the

2. Kierkegaard, *Journals*, xi, A 48; xii, A 320.
3. Théodore de Régnon, S.J., *Étude sur la Sainte Trinité*, vol. iii (1898), p. 458.
4. 1596–1634, the favourite disciple of John of Saint-Samson. Cf. Henri Bremond, *Literary History of Religious Thought in France* (1916 French), vol. ii, p. 287; English trans. 1930.
5. Cf. the expression used by Blessed Henry Suso: 'Bild mit bilden us triben':—'to drive out image with image.' Cf. Bizet, *Henri Suso et le déclin de la Scolastique* (1946), p. 280.

end, enjoying his work aesthetically. The mind is better compared to a *swimmer* who can only keep afloat by moving and
who cleaves a new wave at each stroke.[6] He is for ever brush-
ing aside the representations which are continually re-
forming, knowing full well that they support him, but that
if he were to rest for a single moment he would sink and
perish.

'However far thought may rise, there is always further to
go.'

'If you have understood, then this is not God. If you were
able to understand, then you understood something else
instead of God. If you were able to understand even partially,
then you have deceived yourself with your own thoughts.'[7]

'Whatever is understood by knowledge is limited by the
understanding of the knowledge. . . . If you have reached an
end, then it is not God.'[8]

\*

When we say that God is ineffable, it does not mean that
we cannot say anything about him![9] It does not mean that
there is nothing to say on the subject, or that there is nothing
to be done but keep silence, or that the names which men have
given him are all of them synonymous, and that one can affirm
anything one likes about God, or deny everything about him,
indiscriminately. Nor does it mean that everything that has
been said is only provisionally or pragmatically valuable. The
ineffability of God—and this is what gives it a precise and
eminently positive meaning—is acknowledged *at the end* of the
dialectic. Those who affirm it do not fall into an empty void.

6. St. Bernard, *De consideratione*, V, vii, n. 16 (PL, clxxxii, 798B).
7. St. Augustine, *Sermo* 52, c. vi, n. 16 (PL, xxxviii, 360).
8. St. Augustine, *De civitate Dei*, XII, xviii (PL, xli, 368).
9. St. Anselm, *Monologion*, lxv (PL, clviii, 212B).

On the contrary, their affirmation is the summit of a thought-process rigorously pursued. It does not nullify the results of that effort of thought, and even in its negation it gathers its fruits.

Our ideas about God need, in fact, to be conducted with as much and more order as our ideas on other subjects. Nothing is worse than a premature 'negative theology'. The interplay of affirmation and negation is not a game without rules. The qualities which are affirmed about God—and they are not all affirmed in the same way or on the same grounds—are only identified, as is right, when they transcend themselves and when they cancel one another out. God is not ineffable in the sense of being unintelligible: he is ineffable in the sense of being above everything that can be said of him. He is always above everything which one *must* in fact say of him at first, and which is never simply revoked subsequently: to *deny* is not to *revoke*, for it is always the same God, *semper major*, who impels us first of all to 'affirm' and then to 'deny' in the course of the same movement, of the same advance.

The ineffability of God is only another name for absolute transcendence. Silence comes at the end—not at the beginning.

> Have we said anything, uttered any sound, which is worthy of God? Indeed, I feel that I have said nothing but what I wished to say; yet if I have said anything, it is not what I wished to say. . . . And a sort of battle with words ensues. Since if what is ineffable is what cannot be said, yet what can be called even ineffable is not ineffable. This battle with words is to be prevented by silence rather than stilled by speech. . . .[10]

\*

In the dialectic of the three ways, which gives us access to a human knowledge of God (*affirmatio, seu positio; negatio, seu*

10. St. Augustine, *De doctrina christiana*, I, vi (PL, xxxiv, 21).

*remotio; eminentia, seu transcendentia*), the *via eminentiae* does
not, in the last analysis, follow on the *via negationis*; it demands, inspires and guides it. Although it comes last, the *via
eminentiae* is covertly the first—superior and anterior to the
*via affirmationis* itself. Although it never assumes a definitive
form in the eyes of the intelligence, it is always the light and
the norm, a cloud of light which shows us the path in the
desert of our terrestrial pilgrimage, a hidden power which
excites us to pursue objective knowledge and compels us to
rectify it. . . . That is why we can enter the *via negationis* and
remain in it without fear, once the necessary preliminary
affirmations have been left behind. Understood in this way,
the *via negationis* is only negative in appearance or negative of
appearances. In other words—and more exactly perhaps—
although it is negative and remains negative, it is the very
opposite of *negation*.

Negativity is not negation. A 'negative theology', a theology
which heaps up negations is, nevertheless, not a theology of
*negation*. The moment of negativity which characterises it
does not consist in *calling in question*—any more than the
movement of transcendence or of eminence implied by
theology is a backward movement.

The affirmation, consequently, remains to triumph in its
highest form. It triumphs by negation, which it utilises as the
only means of correcting its own inadequacy. It triumphs in
negation itself, which does not annul it, but compels the
affirmation to transcend itself, which is only the aspect of the
movement of transcendence which can be grasped objectively.[11]
It triumphs every time because it comes first and because
everything necessarily unfolds under its banner and sign. It is
bound to triumph in the end, because in spite of all the indica-
tions to the contrary, that affirmation is, at bottom, mind itself.

11. Cf. St. Thomas, *Contra Gentiles*, I, xxx.

The mind is not, as has been claimed, 'that which denies'; it is that which affirms. The mind is neither revolt nor opposition nor refusal: it is adherence. Negations, revolts, oppositions, all the mind's refusals, in so far as they are well founded, are explained by the demands of that affirmation and adherence. And if those demands are ignored, the mind can no longer be faithful to its laws, and becomes the slave of the natural forces from which it had freed itself and from which it must go on freeing itself, by using those negations, oppositions, revolts and refusals.

✵

An analogy drawn solely from below leads nowhere. However negative the analogy may be said to be, one must, nevertheless, possess in some sense what one is trying to attain indirectly in the form of an aspiration or basic need. If I start from my experience, and using ideas of justice and love, try by analogy and by denying all limitations to qualify God as absolute Justice or Love, I am immediately faced by an alternative. Either the substance of my enterprise consists in a sort of extra-polation, and what is projected into an inaccessible region remains fundamentally homogeneous with the point of departure; and then we are left with a God made in our own image, and the process is typically anthropomorphic. Or else we have within us (and that in the last analysis is what we presuppose) a possibility, a law of transcendence, which makes us posit an absolute justice and love beyond our grasp. That transcending of ourselves is conceivable only if that absolute is already active within us, in some sense, from the start. The analogy must be inverted; what we, in our experience, call justice and love are only what they are because they verify or express something of this aspiration or of this presence.[12]

✵

12. Gabriel Madinier, *Conscience et signification* (1952), pp. 88–89.

. . . We can now realise how far away we are from the supreme
Good, for we can see justice only as freedom from guilt, and
blessedness only as freedom from wretchedness.[13]

\*

In the end we deny everything which, starting from the
creature, we have first affirmed of God. Nothing escapes that
law. There is no conceivable exception to it. But we do not
deny everything at the same stage of the dialectic, nor for the
same reasons, nor in the same way. In fact, if it is true that
there is an unbridgeable abyss between the unique Being of
the Creator and the totality of creatures, it is no less true that
creatures participate in different degrees in the Being of the
Creator. Where analogies are concerned, one cannot treat
those drawn from the realm of the senses on the same plane as
those drawn from the realm of the spirit, or those which are
the fruit of an effort of abstraction on the same plane as those
derived from personal experience. One cannot equate all
forms of anthropomorphism, regardless of whether they spring
from the body or the soul, etc. Furthermore, we must not
merely distinguish, as did Justin Martyr, between two sorts
of divine attributes, those concerning God himself and others
referring to his operations *ad extra*, and treat only the names
of the former as 'divine names'.[14] We must follow the Pseudo-
Denys in distinguishing between what is said of God without
real truth and what is said of him with truth, although it is
subsequently denied with more truth. And in the Augustinian
tradition, we distinguish between the image and the vestige
of God, more or less distant, more or less effaced. Like John
Scotus, we must distinguish between the words which we
apply to God—between those which are *quasi propria* and

13. St. Bernard, *Sermones de dedicatione Ecclesiae*, iv, 5 (PL, clxxxiii, 529).
14. St. Justin, I *Apol.*, i, 3.

others which are *aliena, hoc est translata*. We distinguish,
again, with St. Anselm, the qualities which are not better
than what is not themselves (*non meliores quam non ipsae*) and
those which are better than what is not themselves (*meliores
quam non ipsae*), and we recall that if we are forbidden to assume
that the perfect Being is something which is in some way
bettered by something not itself (*aliquid quo melius sit aliquo
modo non ipsum*), it is equally necessary to admit that he is
really whatever is altogether better than what is not itself
(*quidquid omnino melius est quam non ipsum*).[15] And following all
the Schoolmen, we distinguish finally the 'mixed perfections'
which cannot be found in God himself at all, and the 'simple
perfections' which must be found differently in him than in
us.

Now, the effect of these distinctions is never completely
abolished, as though they had been purely illusory. The
creative essence is more profoundly known, the more it is
sought for among creatures akin to ourselves (*Tanto altius
creatrix essentia cognoscitur, quanto per propinquiorem sibi creaturam
indagatur*)[16]; a principle which is always true. Our negations
bear, in the end, upon everything, though they are always
relative and have their respective bearings. They are not
equivalents. And if we may be allowed a very imperfect image,
they are not identical at the base, though they converge and
meet at the summit. They are involved in a sort of 'levelling
from above'. Their particular meanings remain very diverse,
although taken together they converge upon the necessary
affirmation that God is always *beyond*.

The degrees of participation are real and various—but 'the

15. John Scotus, *De praedestinatione*, ix, 2 (PL, cxxii, 390–391); *De
divisione naturae*, i, 37, 76 (PL, cxxii, 480, 522). St. Anselm, *Monologion*,
xv (PL, clviii, 161–164).

16. Id., *ibid*. (PL, clviii, 212–213).

cause which does not participate is, above all, participation.'[17]
The spiritual creature is like God—but 'God in his total transcendence is unlike anything else.'[18] 'No name can name the
superessential Deity, and no reason bears upon him; the
Deity remains inaccessible and beyond our grasp.'[19]

Yet how shall we complete the process of negation and
proclaim fearlessly that God is always *above all things* if not by
virtue of some requirement anterior even to conceptual
thought; by virtue, that is to say, of a sort of primary, unshakeable affirmation? It is that affirmation which obliges us,
when the moment comes, to deny everything: and so it cannot
be denied. It embraces within itself the truth in all the affirmations which it has made us reject—the truth which cannot be
isolated conceptually because we are unable to conceive
properly what the *modus altior* or *eminentior*, to use St. Thomas's
expression, or the *modus quidam singularis*, to use St. Anselm's,
can be in God. That affirmation is always the soul of our
negations, and if we came to deny it the process of negation
would come to a halt: then, as the result of the fixity of thought,
which amounts to a denial that God is always *above everything*,
we should fall, not into atheism but into idolatry 'attributing
to the image what only belongs to the truth',[20] by the very
fact that we attribute to truth what only belongs to the image.

The power of negation which is in us is therefore not a
negative force: it obliges us always to affirm God without
ever allowing us to stop at anything unworthy of him, and in
doing so reveals itself to reflection as a doubly positive power.
By the principle which sets it in motion, it makes our idea of

17. Pseudo-Dionysius, *The Divine Names*, xii, 4 (PG, iii, 972).
18. Id., *ibid.*, ix (PG, iii, 932). Cf. St. Thomas, *Contra Gentiles*, I, xxix.
19. Id., *ibid.*, xiii, 3 (PG, iii, 981). Cf. St. Thomas, *De Potentia*, q. vii,
a. 5, ad 14m.
20. Nicolas of Cusa, *De docta ignorantia*, i, c. xxvi.

God, beneath its negative form, not simply but eminently positive.

*

We do not know *what God is*. . . . That may mean one of two things. It has a vulgar meaning—that of mere ignorance; and that meaning is to be rejected. But it has a second, particular meaning which refers to God alone; we do not know what God is, but we know what he is not. Or rather, we say that we know what he is because we know what he is not. The last two affirmations are inseparable. In fact, they are identical. Not to know what God is is to know what he is not. And that is a very exalted knowledge. If we refuse to apply any meaning to God which, as such, applies to creatures, we are affirming that God is distinct from all creatures: in a word we proclaim him as God.

> *Deus, qui scitur melius nesciendo.*[21]
> (God is better known by nescience.)

*

When we consider the problem of the knowledge of God as it presents itself to reason, starting from our knowledge of the world, it is more important than ever to distinguish very carefully between the two questions *an est* (whether it is) and *quid est* (what it is).[22] It is no doubt true that one cannot affirm the existence of a being, whatever it may be, without giving at least some sort of definition of it.[23] But strictly speaking, the definition of an essence is one thing and the deter-

---

21. St. Augustine, *De ordine*, II, xvi, n. 44 (PL, xxxii, 1015). John Scotus, *De divisione naturae*, i, 66 (PL, cxxii, 510B). St. Anselm, *Monologion*, xxvi (PL, clviii, 179–180).

22. William of St. Thierry, *Aenigma fidei* (PL, clxxx, 397B).

23. St. Thomas, *In Boetium de Trinitate*, q, vi, a. 3.

mination of a role quite another. In the case of God only the latter, according to the argument, is possible, at least in a positive form, and it is not necessary to know anything more in a positive way in order that God should fulfil the role attributed to him by the argument; it is even essential not to know more.

In other words, if the *proof* of God's existence, starting from the world, is to be valid, and if it is really to be a proof of *God*, it is not, strictly speaking, indispensable to know anything about the divine essence; on the contrary, it seems indispensable not to be able to know anything. For that is the only way in which one can know something about him as distinct from all else. And if one could know something about him, in the same sense in which one knows something of the world, or of the objects in the world, then that essence would in some measure enter into the categories of thought as do the others. It would 'fall under a genus'. From that moment it would form part of this world, and would no longer help us in any way to explain it. We would have to begin everything over again and would be travelling in a circle.[24]

But what a rich matter for reflection intelligence finds in that very contrast between a world which can be known and explained and him without whom the world would be nothing! What an immense difference between vulgar ignorance and that qualified ignorance! What paradoxical knowledge is contained in that rejection of knowledge! How much there is in the void which opens before it, and how great is the light in that obscurity! How can the intelligence fail to feel that its impotence in face of him whom it cannot define is not a sign of insufficiency but of an unbelievable increase? 'That God cannot be measured is what gives me the measure of

24. Similar arguments will be found in A. D. Sertillanges, *Les grandes thèses de la philosophie thomiste*, pp. 48–49 (English trans. p. 53 f.).

him.'[25] And on the other hand, leaving the world and its explanations aside, surely the evidence accumulates in proportion as the negations grow in strength and precision, and are more clearly dictated by a previous unconditioned affirmation whose incomparable vigour cannot be otherwise expressed.

✳

Our concepts have the power to *signify* God truly—and yet strictly speaking, we cannot *seize* God in any one of them; or rather that is how they truly signify him. God would not be God unless he were—not unknowable but—beyond our grasp.[26] He is always above and beyond all that we can say and think of him: *super omnia quae praeter ipsum et concipi possunt ineffabiliter excelsus.*[27]

'The unnameable is the most beautiful of its names and sets it at once above everything else one might be tempted to say about it.'[28]

In the same way, we cannot limit the operations of the mind which really lead us to God to the arguments by which it does so, if these arguments are reduced to a formula or particularised and, by the same token, reduced to some general scheme. But neither may we conclude that our arguments prove nothing or that they are useless, any more than we could, as has just been said, admit that our concepts could not signify

25. Tertullian, *Apologeticus*, xvii (PL, i, 376A).

26. St. Thomas, *De Veritate*, q. ii, a. i, ad 9m. Cf. Etienne Gilson's comments on the Thomist doctrine regarding the perfections of God in *Le Thomisme* (5th ed.), p. 173–174 (English trans., p. 120).

27. Vatican Council, *Constitutio dogmatica de fide catholica*, c. i (Denzinger-Bannwart, *Enchiridion symbolorum . . .*, 11th ed., 1911, No. 1781, p. 473). In addition to the texts already quoted see St. Ambrose, *De fide* I, x, n. 63 (PL, xvi, 543A).

28. St. Albert the Great, *Summa theologiae*, tract. iii, q. xvi, ad 1m.

God in truth. Without our arguments, the basic operation of the mind could not objectify itself and take shape. We should be unable to grasp it. Only, as many philosophers have recognised, the various arguments, viewed in their particular and objective formulations, never do more than express in a rational form, each in its own way, the essential movement of the mind.[29] The hidden Presence which inspires and sustains them is not such that we can disregard the need to convert it into a proof, and that is the proper function of reason. Yet reason can never capture or canalize in its 'ways' more than a fraction of the abundant sap which continuously revitalises the mind and gives it its essential movement.

That is because the hidden pulse of the mind is beyond the grasp of any analysable logical process, just as Being is beyond the grasp of representation.[30] Or, if one prefers, it is prior to it, the 'common root' and the 'hidden spring' of all these processes.[31]

But even these expressions, for all their truth, are inadequate. And in fact he whom we call Being, and whom others undaunted by the paradox, do not hesitate to call 'Negation' or 'Nothing',[32] 'the eternal Nothing', or 'pure Nothingness' —although immediately correcting themselves—cannot, strictly speaking, be represented by the concept of being any more than by any other concept.[33] One may certainly say, and

29. This is what J. Maréchal says very well in the *Nouvelle revue théologique*, 1931, p. 198: 'The diversity of ways or arguments indicates the diversity of starting-points in the real which is immediately accessible to the mind, which rises from them, by a process which remains fundamentally the same, to the transcendent absolute.'

30. Cf. J. Maréchal, *Le point de départ de la métaphysique*, cahier V, p. 183.

31. Ch. de Moré-Pontgibaud, *loc. cit.*, pp. 506–507.

32. Cf. St. Angela of Foligno, *Vita*, iv, n. 72.

33. Cf. E. Gilson, *Le Thomisme* (5th ed.), pp. 153, 155–156 (English trans., pp. 105–108).

sometimes, in order to avoid leading the unprepared mind
into error, one should say, that the names we give to God—
the name of Being in the first place—do represent him in a
way, though very imperfectly. But if we want to observe the
requirements of accuracy to the very end, then we cannot
avoid adding that we are unable to represent the divine essence
to ourselves at all, strictly speaking, and that there is no name
which, applied to God, signifies him 'quidditatively'.[34]
*Nullum est nomen Dei, quod ipsum quidditative significet, seu rep-*
*raesentet et hac ratione merito ineffabilis dicitur.*[35] 'The exclusion
of any definition in regard to God' is valid even to the point
at which it refers to 'the qualification of God as Being'.[36] *Deo*
*quasi ignoto conjungimur.*[37] If definition always involves, to
some extent at least, determination, how could it be possible
to determine him whose being excludes determination?[38]
Let us not be afraid to recognise it, in spite of ontologist
temptations, in spite of our pragmatic instincts or the sneaking
desire for a God who would seem nearer to us. That is one of
the forms which our love of the truth must take, and it is one
of the supports of our adoration. *Deus semper major.* The
imperfection in our representation cannot but fail to involve,
in turn, its negation: *Deus, de quo negationes magis verae sunt*[39]

---

34. Cf. again Gilson, *op. cit.*, p. 156 (English trans., p. 108), 159, n. 2
(English trans., p. 458, n. 51).

35. Suarez, *Tractatus primus, de divina substantia*, II, xxxi, n. 10, which
does not prevent his saying 'nomina, quae nobis imperfecte Deum repraesen-
tant. . . .' (*Opera omnia*, ed. Vivès, vol. i, pp. 184, 185).

36. A. D. Sertillanges, in *St. Thomas d'Aquin, Somme Théologique* (ed. de la
Revue des Jeunes), *Dieu*, vol. ii, p. 383; Cf. Suarez, *Disput.*, xxx, sectio
12, n. 10.

37. St. Thomas Aquinas, *Prima*, q. xii, a. 13, ad 1m.

38. *In I Sent.*; d. 8, q. I, a. 1, ad 4m.; *Contra Gentiles*, III, xlix; *In I Tim.*,
c. iv, lectio 3, in fine.

39. Isaac de Stello, *In Sexagesima sermo*, V (PL, cxciv, 1762c).

(God about whom negations are more true). But once again it must be remembered that the negation invalidates nothing except the limits of the affirmation which preceded it;[40] and once again, in consequence, the fundamentally *quite positive* sense of the necessary negation comes to light.

As for the 'operation of the mind' which results in the affirmation of God, it is not, in the proper sense of the word, an operation—not in the generally accepted sense, that is, or in the first instance. In its first logical moment it is receptivity, a substantial opening of the mind, a welcome which is in the first instance passive. It is only active in a derivative way. Though here, too, the language which we use needs to be corrected or at least carefully checked. 'At every moment we receive a reason which is superior to us.' We participate in a Light which comes from above. Our intelligence does not grasp the Absolute—in a way which is always abstract—without first of all having been grasped by it.[41] That is what is expressed by saying, according to a tradition which goes back to St. Paul, that if we know God, even naturally, it is ultimately through 'the revelation' of God.[42]

40. Jac. Alvarez de Paz, S.J., *De inquisitione pacis*, V, p. i, app. iii, c. i.

41. Fénelon, *Traité de l'existence de Dieu*, Part I, ch. ii, n. 56. Cf. J. Maréchal, *Le point de départ de la métaphysique* (Louvain, 1917): 'Metaphysics is the human science of the absolute. It translates directly the seizure of our intelligence by the absolute. That seizure is not a yoke, something external, but an internal principle of life.' In a note to the above Father A. Hayen, the editor, remarks that Maréchal does not say 'the seizure of the absolute by our intelligence'. That is very significant, and forcibly expresses the idea which we have tried to emphasise at several points. . . . Cf. St. Paul, *Gal.* 4:9: 'But now, after that you have known God, or rather are known by God. . . .' Cf. L. Lavelle, *loc. cit.*, p. 12: There is 'a secret identity, at the heart of thought, between our activity and our passivity'.

42. St. Maximus Confessor, *Capitula theologica et oeconomica*, century I, xxxi (PG, xc, 1093–1096).

'We must be looked at if we are to be enlightened.'[43]
*Signatum est super nos lumen Vultus tui, Domine!*[44]

\*

When we give a thing a name we imagine we have got hold
of it. We imagine that we have got hold of being. Perhaps we
should do better not to flatter ourselves too soon that we can
name God.[45]

*Non ignorabatur Dei nomen—sed plane Deus ignorabatur.*[46]
(He lives apart from the names which are given to him.)

\*

*Quid ergo est Deus? Quod ad universum spectat, finis; quod ad
electionem, salus; quo ad se, Ipse novit*[47] (What therefore is God?
As regards the universe, he is its end; as regards our election,
he is salvation; as regards himself, only he knows).

\*

One must not say: God is not good. He is incomprehensible; but one does better to say: God is Goodness itself,
and it is that Goodness which I cannot understand. One should
not say: God is not the Father, he is the Abyss; one should
say, 'God is the paternal Abyss'.[48]

\*

Even if it were not impossible for the intelligence to grasp
the whole universe and, in one way or another, exhaust its

43. Victor Poucel. Cf. above, ch. i, n. 7–12.
44. *Psalm* 4:7.
45. Cf. St. Gregory of Nyssa, *Contra Eunomium*, xii (PG, xlv, 1108c).
46. St. Hilary.
47. St. Bernard, *De consideratione*, V, xi, n. 24 (PL, clxxxii, 802).
48. Origen, *On Numbers*, hom. xvi, n. 4.

intelligible essence—would it, in fact, be desirable? Surely
it would be quite horrible. Imagine to yourself that there were
no more discoveries to be made and nothing more to wonder
at. Imagine there were no depths to explore! 'Oh the desire
to desire!' Zarathustra exclaims, 'Oh devouring hunger in the
midst of satiety!' And the same sentiment, stripped of its
Promethean romanticism, can be found in the pages of Angelus
Silesius:

> The world is too narrow, the sky too small:
> Where is there room for my soul?

> The knowledge of the Cherubim is not enough for me:
> My desire is to fly high above him, in the Unknown.

Think of the man whose pact with Satan gave him complete
knowledge. From that moment on he was imprisoned in his
own knowledge: 'He spent his days spreading his wings,
longing to explore the luminous spheres so clearly and agon-
isingly revealed to him by his intuition. . . . His heart panted
for the UNKNOWN, because he knew all things.'[49] God
alone is worthy of the intelligence, and God alone can fill it
because he is inexhaustible: 'Surely we have the right not to
see God? . . . And in not knowing him, I recognise him.'[50]

*

> One should not speak much in this life: one can discourse upon
> the world, on matter and on the soul, on rational creatures
> whether good or bad, on judgment, rewards and penalties, and
> on the sufferings of Jesus Christ; but when one undertakes to
> consider God, not in what he has said or done, but in what he is,
> restraint and sobriety is to be commended.[51]

49. Balzac, *Melmoth réconcilié*. Cf. Albert Béguin, *Balzac et la fin de Satan*, in
*Satan*, Études carmélitaines, 1948, pp. 538–547. (English trans.)
50. Claudel, *La Ville*, second version, p. 293. Cf. H. Urs von Balthasar,
*Phénoménologie de la vérité*, the fine chapter on *la vérité comme mystère*, p. 196.
51. St. Gregory of Nazianzus, *Discourse* 27, n. 10 (PG, xxxvi, 25).

I will tell you, my dear friends and disciples, and you my rivals in the love of truth, what happened to me in my search for God.

In the belief that I should soon attain him, I pursued God with indefatigable ardour. I climbed to the mountain top; with Moses, I pierced the clouds; and withdrawing from the material objects among which my spirit had been dissipated, I gave myself as far as possible to recollection. But just as I began to think that I should be able to rest my eyes on God himself, and see him face to face, I found that I could hardly see, in his works, the back he turned towards me; and even that grace I only received hidden in the rock, that is to say in the Word incarnate for our salvation. I therefore learnt that this first and most pure nature was only known to itself, and that it was hidden by a veil, shrouded by the wings of the Cherubim, covering it like the Ark of the Covenant, and that only a tiny ray of its light reaches us. Whoever you may be, that is how you may become theologians (that is to say, contemplate the divinity);—even though you were Moses and the God of Pharaoh, though you were Paul himself, ravished to the third heaven, hearing hidden words; even though you were raised above those great souls and took your place among the Angels and Archangels, since even those celestial (and more than celestial) natures are further below the knowledge of God than they are above terrestrial and corporeal natures.

Let me put it another way: a profane theologian said with great subtlety that it was difficult to know God but quite impossible to express what one thought of him; but as for me, I would rather say that it is impossible to express the greatness of God in words or to give him a name, but still more impossible to understand him.[52]

If my discourse returns to some point about which I have already spoken, you should not be surprised. For I shall continue to say the same thing about the same thing, with that tremor in the voice, in the spirit and in thought, which I feel whenever I

52. *Id. Discourse* 28, n. 3–4 (PG, xxxvi, 29).

talk of God; and I pray that this same blessed and praiseworthy feeling may also be yours.[53]

\*

Should we understand Yahweh's words to Moses to mean: 'I am he who is', or 'I am who I am'? Do we hear the Absolute proclaiming himself or the hidden God remaining silent? Are we presented with a definition or a refusal to define?

Let us leave the scholars to their exegetical discussions— some of them, perhaps, will add further explanations or a subtler gloss.[54] Why should we not retain both meanings? The first interpretation is, perhaps, difficult to justify as it stands, grammatically and historically; and at first sight the second interpretation may sound a little thin considering the solemnity of the occasion. But although they may seem to be opposed are they not, at bottom, very close to one another?[55]

The first formula is full of grandeur. As far as possible, it names God by the name which properly belongs to him, by the name 'which is more rightly his than the name of God itself'.[56] He is! He exists! He is Existence itself! It expresses a 'metaphysical' truth and gives in striking and paradoxical abridgement, an abstract definition of the 'Supreme Being' which sets it apart, while at the same time refusing to assign any limit to it. In a word, it isolates the Absolute of Being and its eternity. *Non est ibi nisi, Est. . . . Quidquid ibi est, non nisi est. . . . Ego, inquit, sum qui sum. Magnum ecce Est, Magnum Est!*[57]

53. *Id. Discourse* 39, n. 10 (PG, xxxvi, 345c).

54. For a thorough discussion of the problem, see A. M. Dubarle, O.P., *La signification du nom de Yahweh* (*Revue des sciences philosophiques et théologiques*, 1951, pp. 3–21).

55. Louis Massignon, *Soyons des Sémites spirituels*, in *Dieu Vivant*, xiv, p. 87.

56. St. Thomas, *Prima*, q. xiii, a. 11, ad 1m. Cf. E. Gilson, *L'esprit de la philosophie médiévale*, 2nd ed. (1944), p. 50, n. 1 (English trans., p. 433 f.).

57. St. Augustine, *In Psalmum* 101, sermo II, n. 10 (PL, xxxvii, 1311).

(There is nothing there but 'he is'. . . . Whatever is there, is nothing but existence. . . . 'I', he says, 'am he who is.' How great, how great is this word).

The second formula is no less precious. It suggests a concrete personality which escapes us. 'I am that which it pleases me to be.' It expresses a solemn and sacred reserve. Without affirming the mystery of Being in itself, it sets it in relief in the simplest and most forceful way. It recalls the 'irreducible distance between anything said about God and the mysterious reality to be expressed'.[58] Thus it vindicates the independence of the Living God. It is the first manifesto against idolatry in thought.[59]

On the one hand, then, there is the perpetual enigma of him who hides himself in his sovereignty: 'Why do you ask me my name?'[60] On the other hand, pure light, radiating in all directions, offering itself without reserve, but too pure for our gaze.

. . . And the two interpretations end by meeting in the idea that 'He who is' cannot be designated and is beyond our reach, a secret at once disturbing and inviolable. *Nomen quod est super omne nomen*.[61]

✳

'Being without a mode of being is also without name.'[62] The mystics have often said so, though there are two distinct ways of understanding their words.

Being 'without mode and without name' does not mean undifferentiated Divinity, an impersonal Principle, an empty Unity or a 'universal Possibility'. . . . It is not the Being who

58. G. Lambert, article quoted above, p. 915.
59. Cf. Romano Guardini, *Le sérieux de l'amour divin*, in *Dieu Vivant*, xi.
60. *Gen*. 33:29; cf. *Judges* 13:18; *Exod*. 33–34.
61. *Phil*. 2:9; *Eph*. 1:21.
62. Henry Suso, *Le Livre de la vérité* (trans. B. Lavaud), vol. iii, p. 120.

in himself (if one could speak in these terms) is without form
and who is falsified by any attempt to conceive him; but on
the contrary—*omnem conceptum excedens ineffabilis Forma*[63] (the
ineffable Form surpassing every concept)—the Being who is
indeterminate, not because of his poverty but because of his
superabundance. Not the Being who presents nothing in
itself graspable because he is as impalpable as Space: but the
mysterious Being, the personal, inviolable Core, the supreme
Condensation. Infinite Being whose infinity is intensive, which
makes him at the same time the perfect Being. He cannot be
named any more than he can be understood—because he is
*above* all names. If it is true that the ultimate secret within
each one of us rests in our personality, God is the hidden
Being *par excellence* because he is *par excellence* the personal
Being. He is he 'from whom all personality derives and takes
its name'.

The 'Supreme Someone'.[64]

❉

Περὶ Θεοῦ, καὶ τἀληθῆ λέγειν, κίνδυνος οὐ μικρός.
Speak about God, but in proper terms—it entails no small
risk.[65]

❉

You are so great and so pure in your perfection, that everything
of mine which infiltrates into the idea I form of you prevents it
from corresponding with you. I pass my life contemplating your
infinity; I see it and cannot doubt it: but as soon as I try to comprehend it, it escapes me; it is no longer the infinite, and I fall
back into the finite. I perceive enough of it to contradict myself

---

63. Nicolas of Cusa, *De docta ignorantia.*
64. Pierre Teilhard de Chardin, *Le phénomène humain* (1955), p. 332.
65. Origen, *In Psalm.* I, 2 (PG, xii, 1080A).

and to correct myself each time I conceive what is less than you; but hardly have I struggled up again, than my own weight drags me down again.[66]

\*

So let human weakness fall down before the Glory of God, and in expounding the works of his mercy let it always find itself unequal to the task. Let us labour with our perceptions, let us face obscurity in our minds, let us be found wanting in our words: it is good that it should be too little for us that we do realise something, even correctly, about God's majesty.[67]

66. Fénelon, *Traité de l'existence de Dieu*, part ii, chap. v, *Eternité*, end.
67. St. Leo the Great. Cf. *Sermo* xxix, c. i.

# 6

## THE SEARCH FOR GOD

WHAT is the philosopher? And what is the mystic? What is the essential difference between them, if we consider their original 'intention' and take them at the root from which they naturally develop?

Should we say that the philosopher makes use of dialectic, where the mystic relies on experience? Does the mystic plumb the depths of Being, while the philosopher tries to discover how thought engenders or expresses it? Could one say that the mystic is concerned with the immediate and the philosopher with mediation?

In fact, dialectic is common to both of them. The only difference is, perhaps, that the one is mainly affective and vital, whereas the other is rational and conceptual. In each case there is experience, though the experience of the philosopher is active and that of the mystic 'passive'. Certainly there is no more typical dialectic than that which attempts to translate the mystical experience—or at least its tendency—into thought. The alternation of opposites is nowhere seen more clearly, more instantaneously so to speak. . . .[1] It has even been maintained that the most obviously dialectical of Plato's dialogues, the *Parmenides*, is at bottom the most mystical.

1. As in those hexagonal designs which, by an optical illusion appear to throw first the black and then the white into prominence. Cf. William of St. Thierry, *Letter to the Brethren of Mont-Dieu*, book II, cap. iii, n. 19 (PL, clxxxiv, 350B–C).

Could it be that the mystic tends to see Being as personal, and the philosopher to conceive it as impersonal?

And yet, to judge by many of the facts, the opposite might equally well be maintained. For as philosophy and mysticism reach the summits of their aspirations, they would appear to transcend this opposition in one way or another.

Perhaps it would be possible to indicate the essential difference between them more nearly by saying that philosophy is above all the search for the *unifying One*, whereas mysticism is the search for—or the attraction of—the *one One*.[2]

The philosopher starts from the need to explain, a need which is, at least virtually, a desire for a total explanation. What he desires is to unify diversity, and at the same time to diversify the one; he requires a system of relationships which embraces everything and makes everything intelligible. His ambition is to comprehend the universe. And if in the course of his search he comes upon God—as he cannot fail to do—it will be as an explanatory principle and a support for the world, as a unifying One. *Res divinae non tractantur a philosophis, nisi prout sunt rerum omnium principia*[3] (The things of God are not treated by philosophers except as the principles of all things). When the philosopher posits the absolute, it is never the absolute absolutely, but 'the absolute in relation to him'.[4]

Philosophy is the work of the reason. It is a 'science'. But God *in himself* is not an object of 'science' to natural man.[5] He can neither be comprehended nor even named. And, as

---

2. We are concerned here only with mysticism in general not with Christian mysticism.

3. St. Thomas, *In Boetium de Trinitate*, q. v, a. 4.

4. Maurice Merleau-Ponty, interpreting Louis Lavelle in *Eloge de la philosophie* (1953), p. 12.

5. We are not forgetting that St. Thomas defined theology as the science of God. But it should also be remembered that theology is not philosophy. Following St. Thomas himself, whose words are too clear and too often

St. Thomas says, we do not know 'what he is': we only know 'the relation of everything else to him'.[6] 'To deduce God from becoming', or from any other aspect of the world, 'does not mean that one rises to a certain direct knowledge of God with the help of becoming; it means penetrating further into the intelligible structure of becoming itself; or, if one prefers, it means knowing God only to the extent to which he is signified by the essential "transcendental relativity" of metaphysical "becoming".'[7] That suffices, in a sense, to define him. In any case, it satisfies the philosopher in the formal sense of that word which we have indicated. As such he does not ask for more:

*Felix qui potuit rerum cognoscere causas!*[8]

But that does not satisfy man. It does not satisfy the spirit. The mystical aspiration is greater, more fundamental and more total than the demands of reason. The mystic reaches out beyond the supreme Cause and the unifying One, which is, so to say, hardly more than a function, and pursues the One itself. He seeks the One in its being and unity. And the least knowledge of that One is worth more in his eyes than the profoundest and most comprehensive knowledge of all else[9]; and for the sake of finding the One, and being united to it, he is prepared to sacrifice the whole universe.[10]

---

repeated to allow of any serious dispute, we maintain that, to the reason of the pure philosopher, 'God, in himself, is not an object of science'—a statement which we maintain in the sense in which St. Thomas meant it, and which the context makes perfectly clear, and not in some vague or general sense, since such a statement in isolation would be equivocal.

6. St. Thomas, *De Potentia*, q. vii, a. 2, ad 11m., and ad 1 m. Cf. E. Gilson, *Le Thomisme*, 5th ed. pp. 150–159 (English trans., pp. 103–110); A. D. Sertillanges, *Somme théologique, Dieu*, vol. ii, p. 383.

7. J. Maréchal.                    8. Virgil, *Georgics*, ii, 490.

9. St. Thomas, *De veritate*, q. x, a. 7, ad 3m.

10. Cf. J. Maritain, *L'expérience mystique naturelle et le vide*, in *Quatre essais sur l'esprit dans sa condition charnelle*, pp. 139–140, 162.

When, as a child, St. Thomas Aquinas exclaimed 'I want to
understand God', it was not so much the budding philosopher
who was speaking as the religious genius, the contemplative,
the potential mystic, the saint with an intellectual cast of
mind. And in so far as his speculation led him to satisfy that
desire, it is not so much the rational science, whose long
career in the West he inaugurated, as one of the aspects of
the 'intelligence of faith' whose ideal and method had been
transmitted to him by the Christian tradition.

But when, on the contrary, he insists so emphatically that
'we do not know God, but only the relations of all things to
him', he is speaking as a pure philosopher. From that point
of view, and that point of view only, there is no reason to
read regret or nostalgia into the phrase. As one of his surest
interpreters writes: 'in a natural theodicy it is not God who
is in question as the subject of science; it is universal being,
the creature. For God is only envisaged and attained as the
first cause and not in himself. In other words, there cannot be
a natural theology apart from general metaphysics.'[11] The
God of the philosophers 'completes the formula of the
world'[12] and fully satisfies their reason.

And yet St. Thomas insists no less that 'the intelligence
naturally desires to know God in himself'. What exactly is
that desire? Does it express the rational need to which
philosophical activity corresponds, or does it, in its own way,
define the mystical impulse in its natural root? Or should one
perhaps see in it the fundamental unity of both?

Let us begin by saying—without for the moment trying to
decide whether the suggestion reveals a faulty analysis or a
deep insight—that, in speaking as he does, St. Thomas merges
the two points of view which we have just distinguished.

11. Sertillanges, *Les grandes thèses* . . . , p. 75 (English trans. p. 83 f.).
12. St. Thomas, *Prima*, q. xxxii, a. 1.

The 'desire to see God', which he regards as natural to us, is certainly, at bottom, mystical in character. It cannot be limited to a desire to comprehend the world. Nevertheless, St. Thomas tries to establish its reality in a purely rational manner, starting from the effects which the intelligence desires to know in their Cause so as to know them fully. With that in view he unfolds a whole argument in the *Contra Gentiles*, which is inspired by his faith, and which a pure philosopher might no doubt criticise as without apodeictic value[13]—and that is precisely why a certain number of his interpreters consider themselves justified in maintaining that the natural desire in question, being the desire to see God *as cause*, is not the desire to *see God* in the full sense of the word.

In brief, the argument consists in showing that human reason, the reason which is responsible for the work of philosophy, is not satisfied with knowing an effect as long as it does not know the cause. Hence that continuous movement, that permanent disquiet, that unrest which lasts until reason, moving from effect to effect and from cause to cause, at last reaches the supreme cause from which everything derives, and which, by that very fact, explains and so unifies everything.[14]

It is a solid argument. But does it, in fact, prove *all* that it sets out to prove? Is the term of the argument *formally* the term envisaged? In its desire to comprehend the universe, the intelligence cannot abandon its search until it has reached the first cause, and one can therefore say, with every show of right, that there is a congenital desire in the intelligence to know that cause.[15] But between that and saying, as St. Thomas

13. Cf. Roland-Gosselin, O.P., *Béatitude et désir naturel* in *Revue des sciences philosophiques et théologiques*, 1929, p. 200.

14. In addition to the well-known texts in the *Contra Gentiles*, see, for example, *Expositio in Matthaeum Evangelistam*, c. v.

15. Cf. Origen, *In Psalmum* II, v. 8 (PG, xii, 1108c).

does in effect, that the intelligence desires to know the first cause, not only as the cause of the effects which it aspires to understand—as the universal *propter quid*—but in its essence,[16] in itself and for itself, independently of its effects and of its relations with everything else, there is surely an abyss?

The mystical impulse, no doubt, bridges the abyss at a single leap. The mystic discerns the One in the Unifying cause, and when he meets the Unifying cause he adheres to the One. But can one say that his strength comes to him from the principle which first moved the intelligence to look for the 'cause of causes'? Could one even say that the mystical impulse simply continues along the path of reason, that it simply goes further in the same direction? Would it not be better to recognise that the philosopher's reasoning conceals an anagogical dialectic, the inspiration of which is quite different from the general desire to know?

St. Thomas, therefore, seems to have failed in his attempt to establish continuity between philosophy and mysticism, between the dynamism of the intelligence and the desire of the spirit. The doctrine of 'the natural desire to see God' is central to his thought, and he has not succeeded in completely unifying it.

No one will succeed where he has failed. The attempt, strictly speaking, is no doubt impossible. The mystical impulse does not exactly prolong metaphysical inquiry; it does not repeat or extend the work, though it can animate it and, in return, be stimulated by it. The root, in each case, is different, the end is different and the basic procedure no less so. Philosophical inquiry rises analytically from effect to cause, in virtue of a rational necessity. The mystical impulse rises from effect, perceived as a sign, to that same cause, by a movement which cannot be wholly justified by pure reason—

16. *Compendium Theologiae*, c. civ.

for if it were an argument, there would be more in the 'conclusion' than in the 'premisses'—but which springs from a need of the spirit no less imperious in its demands than the demands of reason, or more precisely, from the magnetic attraction of Being through its signs. The philosopher may rest from his inquiries in contemplation, once the effect is fully understood; the mystic, in the end, will reject all signs —though he will never quite finish doing so—in order to rest in the contemplation of God alone.

There is, however, something artificial about the distinction originally established. However well-founded it may be, it posits the 'philosopher' and the 'mystic' as abstract beings. It distinguishes two functions of the mind. But while it is true that the functions of the mind are diverse, we must not forget that the spirit is one. The intelligence is steeped in it, and no philosopher worthy of the name would be content to remain for good and all imprisoned in his speciality, even if it were the knowledge and explanation of the whole. Philosophy is always pushing back the frontiers of thought. The philosopher is more than a philosopher, and cannot be reduced to a precise definition. His knowledge of the world is equivalently, or at least becomes inevitably, the perception of his own inadequacy. And the labour of elaborating an intelligible world does not save him from 'the nostalgia of Being'.[17] The greatness of St. Thomas is to have recognised this. By a process which pure reason alone does not suffice to justify, but which the spirit satisfies, or rather insists upon, he was able to penetrate and explore the ways by which the intelligence moves to the point at which he discovered the spiritual appetite within it. In his very philosophy, the philosophical endeavour

17. This is true even of Descartes, so often accused since Pascal wrote against him of only being interested in God for the sake of possessing the world. See in particular the well-known *Letter to Chanut*.

develops into a mystical flight. The human spirit becomes conscious of its total nature and of its high vocation. He explores all its dimensions and, going beyond the techniques and specialisations which obliged him, as it were, to divide himself in two, he seeks to rediscover the simplicity of the mind's essential act.[18] The formal distinctions and oppositions tend to be reabsorbed into unity, although without ever quite reaching it.

The whole of St. Thomas's philosophical research is the search for God.

\*

To reject God because man has corrupted the idea of God, and religion because of the abuse made of it, is the effect of a sort of clearsightedness which is yet blind. For surely the holiest things are inevitably destined to be the victims of the worst abuses. Religion, which is its own source and origin, must continue to purify itself. Moreover, under one form or another man always turns back to adoration. It is not merely his first duty but his deepest need. It is something he cannot extirpate; he can only corrupt it. God is the pole that draws him, and even those who deny him in spite of feeling that attraction, bear witness to him.

\*

God is the Transcendent—but he is also the absolute other. He is the Beyond of the hierarchical universe—but he is equally the unconditioned, the uncoordinated, which no series of conditions brings nearer to us, and which no system of re-

18. Hence the definition of first philosophy which appears to contradict the texts referred to in note 9 above. 'First philosophy is entirely ordered to the knowledge of God as to its final end, and that is why it is called the divine science' (*Contra Gentiles*, III, xxv).

lationships can situate. We can rise up to him—to the threshold of his Mystery—through the 'degrees of being', for he is
the 'being of all beings'—and yet our ascent never really leads
us nearer to him, for if we say that 'he is' then we cannot
really say that other beings are.[19] By comparison with him, all
of them are equally nothing:

> I looked upon the earth, and saw that it was empty;
> I looked into the heavens, and found no light.[20]

The universe is a 'cosmos' whose beautiful order reflects
its Author, the heavens proclaim the Glory of God—and yet
that Glory extinguishes the light of the stars and reduces
everything to dust:[21] silence alone can proclaim it.

Before thy rising Light, everything is a desert![22]

Every creature is an image or vestige of the Creator—
though nothing resembles God.[23] *Similis quidem, sed dispar*[24]
(Similar indeed, but different). *Dissimiles similitudines*[25] (Dis-
similar similitudes). The μή ὄν is incurably opposed to the
ὄντως ὄν—and yet that radical opposition does not exclude
a symbolic relation between that which is not and that which
is; the hiatus makes room for participation. 'The grace and
beauty of creatures are a supreme dis-grace compared with
the Grace of God'—the austere thinker to whom we owe
that uncompromising maxim is also the great poet who sings
of the grace and beauty scattered throughout the creation:

19. St. Catherine of Genoa, *Vita*, xiv. Cf. St. Clement of Rome, *First Epistle to the Corinthians*, xxvii (PG, i, 268).

20. *Jeremia*, 4:23. See the commentary of St. John of the Cross, *Ascent of Mount Carmel*, I, iv.

21. St. Augustine, *Confessions*, XI, iv, n. 6.

22. Paul Claudel, *Vers d'exil*, VI (1912 ed.), p. 236.

23. St. Augustine, *In Psalmum* 85, n. 12 (PL, xxxvii, 1090).

24. St. Bernard, *In Cantica* sermo lxxxi, n. 4 (PL, clxxxiii, 1172D).

25. Pseudo-Dionysius, *Celestial Hierarchy*, ii. 4, 5 (PG, iii, 144A, 145A).

> Scattering a thousand graces,
> He passed through these groves in haste,
> And looking upon them as he went,
> Left them, by his grace alone,
> Clothed in beauty.[26]

The passage from the world to God is thus effected by a double dialectic.[27] On the one hand there is negation, on the other construction. The one suppresses, the other develops.[28] The one is refusal and rejection, the other is acceptance and enhancement.[29] The two movements are interwoven, and neither is altogether independent of the other. We are not faced by a choice between them, and neither has ever brought its task to completion. No mystical ladder reaches its end unless we renounce it. The soul in search of God explores the whole cycle of creation from matter to pure spirit, from the rhythm of the universe to the march of history, but it never passes from one stage of its ascension to the next except by a series of rejections and denials, for the beings which it questions on the road all reply: 'We are not the God you are seeking.'[30]

*

Let us admit that three-quarters, and perhaps more, of all that man says and thinks of God in his worship and his prayer is infected with hypocrisy and superstition, childishness, convention and routine repetitions.[31] Yet we must be on our

---

26. St. John of the Cross, *Ascent of Mount Carmel*, I, iv.
27. Corresponding with this double dialectic there are two spiritual ways, the way of signs and the way without signs. See J. Monchanin, *De l'esthétique à la mystique* (1955), pp. 105–112.
28. St. Augustine, *De vera religione*, c. xxix, n. 52 (PL, xxxiv, 145).
29. St. Anselm, *Proslogion*, c. xiv (PL, clviii, 234D).
30. St. Augustine, *Confessions*, X, vi, n. 9.
31. Julian Green, *Journal*, 30 June 1943 (vol. iv, p. 54).

guard against contemptuous judgments, because they are the most blinding of all. This enormous wastage must not blind us to the spark of truth that burns in the innermost recesses of the soul. Even when it conceals it from us, it is not always stifled, and from time to time it can be seen glowing and bursting into a pure and upright flame.

\*

*Optimi corruptio pessima.*

The coating of hypocrisy is never so thick as round the idea of God.

\*

God can never really be thought or recognised apart from a *sursum*, which no proof can ever arouse. It is much less important to prove God to the unbeliever than to open his eyes. Apologetics is to testimony what the sermon is to example.

\*

If, when night comes, I think back to certain privileged moments when the truth of my affirmation was revealed to me in an experience, I am not living on a deceptive memory, on the recollection of a pleasing experience, but recollecting a value perceived; it is not the recollection of the fulfilment of a value which I bore in principle within me, but the recollection of a newly discovered existence which integrates, orders and judges all human values.

I had been told that the grey canopy of the sky was only a thin curtain of cloud which hid the sun. I had been offered ingenious and even convincing proofs. They explained many things. That fine solution was a correct one. My reason had nothing more to say. And yet its direction was not unalterably fixed. My mind remained perplexed. . . . One day the clouds opened, and I saw the sun appear beyond them. I was unable to fix my eyes upon it, but I was struck by its rays. My countenance was illuminated.[32] From then on the trial was no longer a scandal. The clouds are once again opaque, but they cannot make me doubt the sun.

Perhaps, if I get caught up in a network of argument, it will be enough to meet a man for whom the clouds have in effect opened. Perhaps it will be enough to see a man who has seen, and to believe on his testimony. For that is the miracle which is endlessly repeated, generation after generation, which overcomes our prejudices and all the precautions we oppose to it: it blows a breach in the critical fortress and dynamites negation. Such a testimony is unlike any other we encounter in ordinary life. Through his testimony, through the man who has seen, I really see—or at least glimpse or have an inkling of what he has seen. The sound of his voice awakens an echo in me. The night in which I live is illuminated, without ceasing to be darkness. And what the psalmist says to God I can say to the 'man of God': *in lumine tuo videbimus lumen* (in thy light we shall see light).

The saints are the efficacious witnesses of God among us.[33]

\*

When we meet a saint we are not discovering at long last an ideal, lived and realised, which had already been formed within us. A saint is not the perfection of humanity—or of the superman—incarnate in a particular man. The marvel is of a different order. What we find is a new life, a new sphere of existence, with unsuspected depths—but also with a resonance hitherto unknown to us and now at last revealed. We are shown a new country, a home we had originally ignored, and as soon as we perceive it we recognise it as older and truer

32. St. John of the Cross, *Spiritual Canticle*, xiv, n. 20.
33. See Léonce de Grandmaison, S.J., *La religion personnelle*, pp. 177–179.

than anything we had known and with claims upon our heart.

No feeling of self-satisfaction invades us; we do not see our noblest image reflected in a mirror. This is not the fulfilment of our loveliest dream—or rather there is something further, which is not only more beautiful: we are simultaneously attracted and repelled, and the more we are repelled the more we are attracted. We experience an ambiguous sensation as of something at the same time very near and very far; something disturbing, troubling and at the same time obscurely desired. The feeling is a mixed one, compounded of a sense of strangeness and of supreme fulfilment beyond all desire. We are both disconcerted and ravished, and the delight we experience is never without a sense of dread. Our worldliness reacts to the threat. Our secret connivance with evil is aroused. We are on our guard. If we had begun to regard ourselves as perfect in some respect, we shall be doubly tempted to reject the provoking vista which is going to oblige us to recognise our misery and, more than that, the wretchedness of what we call perfection.

But in all this we are not left to ourselves, as spectators. It acts upon us as a provocation. It is a summons to choose and to act, unveiling our most hidden tendencies. . . . All of a sudden the universe seems different; it is the stage of a vast drama, and we, at its heart, are compelled to play our part.

If there were more saints in the world, the spiritual struggle would only be more intense. As the Kingdom of God becomes more manifest, it calls forth more fervent adherents—and, correspondingly, more violent opposition. The heightened urgency of the situation provokes tension and becomes the occasion of resounding conflicts.

For if we are more or less at peace in the world, it is simply that we are tepid.

*

'Love and do as you will', St. Augustine said—if you love enough to act, in every circumstance, according to the dictates of love.

One might also say 'love and believe what you will'—if you know how to extract all the light from love, whose source is not in you.

But do not rush to the conclusion that you know what love is.

*

If the task of reason is to penetrate sensible appearances, the task of faith is to penetrate all appearances. It must pass through all the nights. That is what sometimes makes it so hard; it is the very opposite of a 'lazy solution'.

Faith is always a victory.

*

The solitary mystic sees himself as identical with the Principle of Being, and so infinitely increases his solitude; the believer is brought up short against the Other, is overthrown and, after the struggle, united with him in love.

*

When the witness of the saints incites my adherence, I do not confuse the power of their testimony with the force of a rational argument. I know perfectly well that I am not effecting a scientific operation. I can see quite clearly that there are two *genera*, and that their difference cannot be bridged. But although their testimony is not a proof, that does not mean that it is a bad or even a weak proof, any more than it is an apodeictic one. And so I shall not say that it is 'reasonable and prudent' to rely on what I am told by men who deserve respect, whose affirmations converge and whose sincerity is beyond doubt, although the evidence which they claim to bring remains 'purely extrinsic' and does not allow me to draw any conclusion with real certainty. I do not need to be told that such an argument is 'devoid of scientific value', since I have already admitted that there is no question of arguing. What I contest is that the testimony to which I give my adherence is 'purely extrinsic'. On the contrary, its whole value—which, to repeat, has nothing 'scientific' about it—consists in the echo it evokes in me. It enables me to unravel something essential within myself. That does not mean that it supplies a proof. None is even hinted at. But the two epithets 'reasonable' and 'prudent' are none the less inadequate to describe the adherence which that testimony compels me to give.[34]

The witness of the saints does not produce an automatic effect. Nor can it be generalised in the same way as a rational proof. But when it is efficacious, it is an altogether different thing and not a simple and inferior form of proof.[35]

\*

'Once a thing is explained it ceases to interest us' (Nietzsche). So God interests us eternally—and everything else in God through its participation in his infinity.

In the 'now' of eternity, everything will be 'new, fresh and present' to us in God.[36]

\*

To some people God is the one who lets them sleep in peace, a reassuring word which dispenses them from the fatigue

34. J. Maritain, *Approches de Dieu*, p. 118 (English trans., p. 107 f.).
35. Cf. Fernand van Steenberghen, *Le problème philosophique de l'existence de Dieu* in *Revue philosophique de Louvain*, vol. xlv (1947), pp. 146, 302.
36. Master Eckhart.

of inquiry. To others he is the one who tears them from the 'false security' in which, according to Pascal, the world lived before the coming of Christ.

*

The humility of the saints is not the humility we attribute to them. Nor is their love what we imagine it to be. And to say everything—if we must—our God is not their God.

Yet each one of us, at the bottom of his heart, has some inkling of the difference, and can begin to measure the gulf. And that knowledge helps us to reduce it. Each one of us, if he will but attend, can have some premonition of the strange new country in which the saint finds his home.

*

Hell is the work of man, of the man who refuses to give himself and puts himself in bondage: to whom love is unbearable.

As the first Christians understood so well and symbolised so admirably, it is one and the same gesture which both saves and condemns, the serene and majestic gesture with which Christ shows the five wounds. The Redeemer does not transform himself into the Judge, as though tired of his first role; it is the same unique love, the same unchanged love which pronounces the double sentence as it is refracted in our hearts.[37]

It is the same word, the one double-edged sword which comes to some as the word of life and to others as the word of death. *Semel locutus est Deus, duo haec audivi* (God has spoken, and these two things have I heard).

37. See, among others, the fresco of the Last Judgment by Cavallini at the farther end of the tribune in the Church of St. Cecilia in Rome.

It is the same 'contemplation' which is obscure or luminous, exquisite or cruel, according to the state of the subject.[38]

In its unchanging essence the same divine Fire is pain for one, Purification for another, and Beatitude for a third.[39]

\*

*Noverim me, noverim te.* May I know myself, and may I know you, O God, my God!—only that double wish must not be realised at two different times. I cannot get to know myself without seeking to know God—for in my very being I am wholly relative to God. The subtlest investigations and the most learned reflections only serve to lead me astray instead of revealing me to myself so long as I try to know myself alone. Man only knows himself—can only desire and love himself— in God or before God. *Noverim te, noverim me.*

The man of prayer discovers in himself and upon himself the light which the man in search of his 'self' does not discover.

\*

Man, alas, is above all frightened of God. He is afraid of being burned at his touch, like the Israelites who touched the Ark. That adds subtlety to his denials, cunning to his attempted escapes, and makes the pious inventive in devotional tricks to deaden the shock. . . . Whether incredulous, indifferent or believers, we compete with one another in ingeniously guarding ourselves against God.[40]

\*

38. St. John of the Cross, *The Dark Night*, chap. vi.
39. Alexander of Hales, *In I Sent.*, d. 37, n. 10.
40. Cf. Simone Weil, *Gravity and Grace*, p. 52. 'We fly from the inner void since God might steal into it.'

'Whenever we seem to touch upon God, or when we perceive that he has come to us in our dreams and our wretchedness, we are horrified.' (Maurice Blondel).

\*

The mystical impulse is not a luxury. Without it the moral life would run the risk of becoming a form of repression, asceticism a withering dryness, docility a form of sleep and religious practices a routine, a matter of display if not of fear.

\*

The genuine mystic confides in no one—not because he is prudent or aloof, nor solely from humility or love of mystery. He has no confidences to make. The life of consciousness is beyond the range of psychology, particularly in its highest form, the mystical life.

\*

The mystic longs to know God in himself; that is to say, as God knows himself and—if love has revealed itself to him in some measure—he longs to love God for himself, that is to say with the love with which God loves himself. And then he will be open on all sides to the inflow of the divine.

\*

*Cum absens putatur, videtur;—cum praesens est, non videtur*[41] (When he is thought to be absent, he is seen—when he is present he is not seen).

\*

Is mysticism 'an intuition of God'? Yes, but always in the

41. St. Augustine, *De videndo Deo (Epist. 147 ad Paulinam)*, vi, n. 18 (PL, xxxiii, 604).

dark night. For God is only found by always seeking him. He
is always 'the one sought'.[42]

*O Luce qui mortalibus*
*Lates inaccessa, Deus.*[43]

(O God who dwelleth in light which is for mortals inaccessible!)

✳

Progress in the knowledge of God there is, but it would
not merit the name of progress unless it took us nearer the
term, and in another sense left us just as far away. As the
infinite allows us to approach it, so it proves the more inaccessible. Moreover, 'those who ascend never cease ascending';
those who have started on their course move 'from beginnings
to beginnings, through beginnings without end'.

'Once the soul takes flight and begins, in so far as it can, to
participate in the divine good, then the Word begins to draw
it to itself as though it were still at the beginning of its
ascent. . . . "Arise" it says to the soul which has already
arisen, "Come" it says to the soul which has already come.
And indeed those who really rise up must always continue to
do so, and those who run towards the Lord will never find
their journey to the divine cut short. In saying "Arise and
come", the Word obliges us to rise up continually, and never
to slacken speed, always giving us grace for a new and more
perfect ascent.'[44]

Far from discouraging us, that thought can only strike us
with wonder. What it teaches us has nothing to do with
Penelope's web or the rock of Sisyphus. Nothing is ever lost,
the distance we cover is not wasted, and there is no turning
back upon our steps—but everything is greater and more

42. St. Gregory of Nyssa, *In Cantica canticorum* (PG, xliv, 720c).
43. Gallican liturgy, hymn for vespers.
44. St. Gregory of Nyssa, *In Cantica canticorum*, hom. v. (PG, xliv, 873–876).

beautiful than we could have imagined or suspected. For God must always be greater than everything 'not only in this world but in the next'.[45] Everything, therefore, which has to do with God always preserves the freshness of a beginning and the zest of the original departure. No fatigue or satiety 'which would dull the spirit'[46] need be feared. The rich autumn harvest will have the savour of the first fruits of spring. And we ourselves shall participate in this eternal youth. We shall understand more and more as we experience it, and as we see better and better that we do not yet understand it, and never shall understand it, what this astounding thing, the discovery of God, means—for it will never cease to astonish us.

> *Cum consummaverit homo, tunc incipit.*
> *Sanctorum sicut aquilae juventus renovabitur.*[47]
> (When man is consummated, then he really begins.
> The youth of the saints shall be renewed like eagles.)

*

'Not to be able to reach God is our discovery; the failure itself, our success.'[48]

*

'God is not a spectacle. The contemplation of him is something more secret, veiled and disconcerting. He is only discovered, and then only in a certain degree, in the fidelity of our movement towards him, in a "passover" which brings peace out of suffering and gives riches at the cost of stripping ourselves of everything.'[49]

*

45. St. Irenaeus, *Adv. Haer.*, ii, 28, 3 (PG, vii, 806A).
46. Leibnitz, *Principes de la nature et de la grâce*, n. 18. Cf. Gratry, *Connaissance de l'âme*, vol. i, i, 13.
47. *Psalm.* 102 (103): 5.
48. Eckhart, *Treaties*, xiv.
49. Jacques Paliard, *Profondeur de l'âme* (1953), p. 159.

. . . It should not be imagined that the soul always remains or should remain at the highest point of the spirit, and so adheres to God as a most pure spirit, in whose presence all things are as nothing . . . in such a way that progress consists solely in inserting, immersing and concealing itself deeper and deeper in the divine spirit. Spiritual profit should be conceived in this way: once the soul has reached the summit in one order or degree, then if it is to be raised by God to another degree, substantially more perfect, it must first of all return to the lower state and begin a new and more searching purgation, an expansion and a fresh disposition, a deeper and more sincere foundation of true self-knowledge than heretofore. Which new beginning, nevertheless, virtually contains in its lowliness all the heights previously attained.

For this must be carefully noted: all that the soul acquired at the summit of its spirit, the sublime fruit and final term, she now possesses in secret by way of principle, by way of being, as basis and substance, hidden and unknown, as something which is joined and identified with her own substantial being in this new beginning at the lower levels: and this, by a singular disposition and artifice of God, in order that the soul should not esteem its interior state too greatly and magnify it; so that being exempt from the danger, she may continue always to grow in God. . . .

Hence it comes that true spirituality does not always consist in enjoying God; nor, similarly, in always being able to persevere continuously at the summit of the spirit; but in being able to follow God according to all the changes and vicissitudes and fruitions and all the diversity of degrees which he imposes upon the soul. In a word, the ability to follow him wherever he leads.[50]

\*

The real problem is not 'the search for God'—for there are ways of searching for him which are no more than provoca-

50. Constantin de Barbanson, *Anatomie de l'âme et des opérations divines en icelle* (1635), Part iii, art. xvi, pp. 155–158.

tions[51]—and any search in which man allots himself the principal role is surely a provocation. The real problem is to cultivate the right dispositions so that one may hope to find him without—so to say—having to search for him. The essential thing is to understand that these dispositions themselves can only come from God. For it is he who searches for us and who, in his time, will manifest himself to us.

> Turn towards the East and await God,
> And the dawn of grace will soon rise in you.[52]

\*

Sometimes we think we are looking for God. But it is always God who is looking for us, and he often allows himself to be found by those who are not looking for him.

\*

No critical ingenuity can ever prevail over the clearsightedness of the pure in heart.

The pure in heart are twice blessed: they shall see God and through them God will make himself seen.

\*

That which man, starting from his own level, calls 'God' is a vital impulse, the topmost summit of the world whence the summons leads into the beyond, a jumping-off place for the leap of faith. But if one stops at that point, giving it a definitive value, the result is a 'religious' deception, and nature or some ideal or other is turned into God. Such a divinity, starting with the *numina* of natural religions, and ending with the absolute being of religious philosophy, has no real existence. The God who is,

51. Cf. *Wisdom*, 1 : 2.
52. Angelus Silesius, *The Cherubic Pilgrim*, ii, 215.

the true and living God, is he who shows himself to us in Revelation. It is with him, whether man likes it or not, that he is concerned in time and in eternity.[53]

*

Light is the Lord's cloak; rest assured that if you lose the light you have not yet lost God himself.[54]

*

Beyond all conventions—in the rejection of all untruth—at the cost of security—behind all negations—when everything fails—in the abandonment of everything:

The discovery of God.

*

What is an unhealthy subtlety or at least a superfluous refinement to some is a necessity to others. It is the 'flight ahead' to which they are condemned. That narrow defile is their only path to salvation.

They are certain to disconcert or antagonise the easily satisfied mind, more than the clever and the restless, more even than the adventurers in the world of thought with their dubious designs, more than the disguised adversaries of the truth. But how can they help it?

Their choice is between scepticism and the purification of faith. Between despair and the purification of hope. Between hatred of their kind and rebellion, and the purification of love.

53. Romano Guardini. It should, however, be noted that the 'natural' God whom the author excludes is only the one in whose name one would exclude the God of supernatural revelation by attributing to him 'a definitive value'.
54. Angelus Silesius, *The Cherubic Pilgrim*, ii, 5.

And the Peace which comes upon them co-exists with an anxious dread.[55]

And the God of their distress is more *God* to them than any other—and is *theirs* more than any other.

And no other God is so contagious—and no distress so effectively appeases the mind without lulling it to sleep.

✻

No, my Love, you are neither fire, nor water, nor aught that we say. You are what you are in your glorious eternity. You are: that is your essence and your name. You are life, divine life, living life, unifying life. You are all beatitude. You are ineffable, incomprehensible unity, supremely adorable. In a word, you are Love, and my Love. What, then, shall I say of you? You made me for you; for you, I say, who are Love. Why, then, should I not talk of love? But alas, what can I say? On earth I cannot speak of it. The saints who see you in heaven adore you in silence, and their silence is a sacred language in which they taste love. You pour your love into us, O my God, as into them. And you fill us with yourself, as you do them. Why, then, should we not do as they do? Why should we not taste love as they do? For if you are their Love, you are also ours. They see you directly, O my dear life, and that is what they have and what we have not in the lowliness and the misery of the flesh. But when we are delivered from our prison, we shall see you as they do, we shall praise you as they do, we shall embrace you as they do, we shall possess you as they do, we shall be immersed in you as they are, and we shall no longer express your love in humble similitudes, because we shall be nothing but love, because we shall be wholly in love, that is, in you, who are my one Love, my mercy and my All.[56]

✻

55. Cf. Marie de l'Incarnation, *Relation de 1654*, xii. There are, however, different forms of anxiety. Cf. H. Urs von Balthasar, *Der Christ und der Augst*.

56. Marie de l'Incarnation, *Exclamations et Elévations*, ii.

Let nothing disturb you,
Let nothing frighten you,
> Though all things pass,
God does not change.

Patience wins all things,
But he lacks nothing,
Who possesses God:
> For God alone suffices.[57]

✻

But if I do not reach my goal? If I fall by the wayside? I shall nevertheless have the joy of having run, strained and sweated as much as I could, in search of the face of my Lord.[58]

✻

My Lord and my God, my one hope, hear me, lest in my lassitude I no longer wish to seek you, but let me always search for you ardently. Give me strength to seek for you, you who have let me find you and have given me the hope that I may find you more and more. Before you is my strength and my weakness: hold up my strength and heal my weakness. Before you is my knowledge and my ignorance: welcome me when I try to enter where you have opened to me; and when you have closed the door, open to me when I knock. Let it be you that I remember, that I understand, and that I love! Let those three gifts be increased in me, until you have reformed me entirely. . . . Deliver me, Lord, from the abundance of words from which I suffer inwardly, for my soul is all misery before your gaze and takes refuge in your mercy. For my thoughts do not remain silent even when my mouth does not speak. . . . Innumerable are my thoughts, such as you know them, the thoughts of a man, for they are vain. Grant that I may not consent to them . . . so that I do not wallow in them as in a sort of

57. St. Teresa, *Poems.*
58. Richard of Saint-Victor, *De Trinitate*, III, 1 (PL, cxcvi, 915–916).

sleep. . . . 'We shall say much, and yet shall want words: but the sum of our words is, He is all.' When we have attained you, the words which we multiply without attaining you will cease: you alone will remain everything in all of us and we shall only say one single word, praising you with one single movement, and making one single whole with you. . . .[59]

✳

*O ergo, quem nemo quaerit vere et non invenit, quippe cum ipsa veritas te quaerendi in conscientia quaerentis non suspectum jam habeat responsum aliquatenus inventae veritatis![60]*

(So no one truly seeks you without finding you because the very truth that you are being sought in the consciousness of the seeker contains in itself the unsuspected answer of a truth which, in a measure, has been discovered already.)

✳

To await God is to possess him.[61]

59. St. Augustine, *De Trinitate*, XV, xxviii, n. 51 (PL, xlii, 1098).
60. William of St. Thierry, *Speculum fidei* (PL, clxxx, 397A).
61. Fénelon, *Œuvres* (Paris ed.), vol. viii, p. 557.

# 7

## GOD IN OUR TIME

WHENEVER it abandons a system of thought, humanity imagines it has lost God.

The God of 'classical ontology' is dead, you say? It may be so; but it does not worry me overmuch. I have no inclination to defend the petrified constructions of Wolf. And if 'classical ontology' disappeared, it was surely because it did not correspond adequately with being. Nor was its idea of God adequate for God. The mind is alive, and so is the God who makes himself known to it.

'God is dead!' or so at least it seems to us . . . until, round the next bend in the road, 'we find him again, alive'. Once again he makes himself known, in spite of all that we have left behind on the road, all that was only a viaticum for one stage of our journey, all that was only a temporary shelter till we had to make a fresh start. . . . And if we have really progressed along the road, we shall find God himself greater still. But it will be the same God. *Deus semper major*. And once again we shall move on in his light.

God is never left behind among the dross. . . . In whatever direction we go, he is there before us, calling to us and coming to meet us. . . .

\*

It is only too true, often enough 'a deist is a man who has not had time to become an atheist'.[1]

The deist's God, the God of several modern 'theodicies' which weigh and measure him rather than defend him, the God who can hardly say 'I am' any longer, the God who tends to be no more than 'the universal harmony of things', who rules over a beyond where 'everything is the same as here',[2] the God imprisoned 'within the limits of reason', who no longer intervenes in the world, who is really nothing but the projection of natural man, who is distant yet without mystery, a God made to our measure and defined according to our rules, a God merged in the 'moral order of the universe' as man understands it, a God who is not adored and whom one can only serve by the cult of morality, a God who is 'only accessible in pure knowledge' and who is 'nothing but that knowledge itself', a God in fact whose thoughts are our thoughts and whose ways are our ways: such a God has proved very useless in practice and has become the object of a justified *ressentiment*. And when at last man decided to get rid of him altogether in order to enter into his own inheritance, he was only a shade, 'reduced to the narrow limits of human thought'.[3]

\*

1. Bonald. Proudhon put it quite crudely in his *Philosophie de la misère*: 'I know a man who would be ready to draw his sword in God's cause, and, like Robespierre, to set the guillotine going till the last atheist had been destroyed, little suspecting that he would be the last.'
2. Cf. Jules Lachelier, criticising Leibnitz's conception, *Lettre à Jean Baruzi*, 10th December 1906.
3. It should, however, be remembered that each case is unique; that something active very often survives; that there are ambivalences, that the 'quarrels over atheism' are often muddled, and that what looks like a degradation in one light may be sometimes in fact the beginning of a rediscovery.

'To convict Voltaire of atheism is not really a great victory over Christian thought'[4]—nor is it, for that matter, to show that the God of Fichte or Hegel easily turns into the Man of Feuerbach. 'Let them reach what conclusions they please against deism,' Pascal prophetically remarked.[5]

\*

We have witnessed, during the last few centuries, 'the rationalistic evaporation of God'[6]. But it was the rationalist God. A single puff will disperse the vapour. We shall not be disturbed. We shall even breathe more comfortably. The true God, the God we continue to adore, is elsewhere. He is everywhere you think to find him. He is everywhere, even when you do not find him.

\*

When 'God's cause' is lost, then God is victorious once again. Then 'he is his own defender'.[7]

\*

It is generally conceded that Christianity 'inaugurated the struggle against false gods'. But some people would like to take over from it, as though it could not complete the task itself. They would like to make philosophy the heir to Christianity. Yet the philosopher, it is said, is 'the man who understands' and not 'the man who chooses'.[8] In that case the false gods still have a promising future!

\*

4. E. Gilson, replying to Brunschvicg in *Querelle de l'athéisme*.
5. Cf. G. Fessard, *La main tendue* (1937), pp. 124–126.
6. G. Gusdorf, *Mythe et Métaphysique* (1953), p. 221.
7. Leibnitz, *Causa Dei* . . .
8. Merleau-Ponty, *Eloge de la Philosophie* (1953), p. 65.

One must 'reject the gods', a certain writer says, 'all the gods'. That is precisely what the disciples of Jesus taught us to do from the beginning. If they were taken for atheists, it was not because they were making the banal claim to have discovered *another god*—who would simply have been one among many: but because they proclaimed him who is *totally different from the gods*, and who frees us from their tyranny. They denied everything that the men around them took for the divine—everything that man, at every epoch, tends to deify in order to adore himself and tyrannise over himself, in and through his gods.

The Gospel is the only 'twilight of the gods'.

\*

It is possible to maintain that religion—faith in God in the first instance—is a system invented by nature with the object of *reassuring* man who would otherwise be paralysed by fear in face of a hostile mystery.

But there is another way in which man can reassure himself: the rationalist way, the way of the short-sighted optimist who does not even rise to the level at which the mystery can be felt and proudly announces that there is none to know.

Which of the two is nearer the truth?

Faith in God certainly gives us confidence.[9] That is undeniable, and there is no reason to be ashamed of it, as though it were more intelligent not to have been touched by dread or anguish, or nobler not to wish to be delivered from it. Faith, indeed, reassures us—but not on our level, or so as to produce a paralysing illusion, or a complacent satisfaction, but so as to

9. Cf. Clement of Alexandria, *Texts chosen from the Prophets*, n. 21 and 26: God is at the same time 'inaccessible light' and 'devouring fire'; as fire he engenders fear, as light he gives back security.

enable us to act. It gives man the confidence to become worthy of himself, and helps him not to succumb in the great crisis in his growth to maturity, when consciousness awakens from animality. Faith gives him confidence, but it does so by establishing him in the truth and by communicating a disquiet of a higher order.

*

What could be more horrifying than a world without God, without stability, and without mystery, convinced of its own transparency, falling headlong into an abyss of meaningless and endless change, *dum nil perenne cogitat*, while the soul thinks of nothing eternal? or a society entirely given over to temporal idols, in which the *mens avida aeternitatis* is suffocated to death —a world of inexpressible horror and despair?

*

To compare Nietzsche to Jesus: Jesus was killed because he proclaimed the Father who is in heaven; Nietzsche killed himself, his mind foundered in perpetual night, because he proclaimed, accepted and willed 'the death of God'.

Since that decision was taken, in spite of deliberately persuading himself that he possesses a 'carefree knowledge', man is obliged to admit, with Nietzsche, that his knowledge leaves him 'frozen stiff with fear'—he is a prey to 'a sacred terror'.[10]

*

The divine right of kings, the divine right of peoples: both of them human inventions and instruments of oppression. The divine right of God is the only source of freedom.

*

10. Cf. my *Affrontements mystiques* (1950), chap. iii, 'Nietzsche mystique'.

The anti-theist—the militant atheist—claims to know God, otherwise he could not oppose him. But by that very fact, and whatever he may say, he is not really opposing God. For God cannot be known in that way.

\*

Even from the point of view of sociological analysis, the Marxian theory of religion is hardly exact—or at least it is incomplete. Religion, let us admit for argument's sake, might really be the opium of the people if the people had that particular craving. In certain favoured circumstances, perhaps, they have. But observation suggests that as a people becomes a proletariat, it loses that taste. Far from stimulating the religious impulse by a sort of mystical compensation, the increasing 'alienation' and isolation which goes with the proletarian condition tends, on the contrary, to smother any interest in religion. It turns those whom it dehumanises away from God.

For in fact a certain degree of social 'alienation' very often involves the alienation of the consciousness. And the alienated consciousness is the exact opposite of the religious consciousness.

\*

'The proletarian has no country.' In an analogous sense and for similar reasons, the proletarian has no religion. In a society such as ours, religion tends to become a luxury article, which is denied to a whole section of the population. Though the suppression of the proletariat will not automatically give God back to man : but it is, up to a point, a condition of God's being given back to him.

\*

The famous Marxian dialectic is another perfectly authentic, though very heterodox and shameful, example of a 'Christian philosophy'.

Hegel was a theologian; his main categories were supplied to him by a rationalisation of the Christian mysteries. Marx had been a Hegelian. It took a double inversion, a double 'apostasy', to produce the final result: the divine was made immanent by Hegel; then the human was materialised by Marx (what Marx called standing the dialectic on its feet again). But in spite of its double metamorphosis, the texture remains unchanged: it was furnished neither by experience, nor by science, nor even by pure reflection (does pure reflection yield anything?) but by faith. It is the Mystery of Christ, God incarnate, dead and risen. A keen eye can still detect the 'theft of sacred things' at the source of categories which are in appearance the most profane. Even now, after so many changes, so many denials, negations and corruptions, the Marxist ideology is still really a parasite that draws life from the Christian substance.[11]

Long after people have stopped talking about 'the absolute history of the divine idea' or its 'supreme alienation', faith in the God who is incarnate in history, and who was 'destroyed' for our sake, will still be alive. When the 'Good Friday of Speculation' and the 'Calvary of History' are long forgotten expressions, the Cross of Jesus will still bear the Fruit of Life.

✶

If I meet a saint, I know what I have seen—or at least glimpsed. But some people say they can do without God from

11. Cf. Franz Grégoire, *Aux sources de la pensée de Marx* (1947), p. 77; G. van der Leeuw, *L'homme primitif et la religion*, pp. 194–195. Edgar Quinet in *Le Génie des Religions* (1841) had noted the same phenomenon.

now on, and can find something better. I wait for them to show me a new type of saint.

*

To think that people can convince themselves that 'metaphysical anxiety' is a thing of the past! 'We are cured of our obsession' they tell us, 'cured of our folly: of our obsession with God, with being and with nothingness, of the searing burn of the unknown in the heart of the known, and of *the other* whom we pursued in our dreams.' We are no longer 'haunted by the absolute' they tell us, for we have shaken off the burden of 'eternal truths'. . . . Poor mutilated wretches who think they have achieved freedom, and celebrate the most lamentable abdication as a 'tremendous victory'. They had better sing their hymn of victory while there is time. For even in them the mutilation is not final and they do not realise that man cannot abdicate. A sudden awakening can put everything in doubt, and a single spark can relight the fire that seemed to have died out. The soul comes to life again though we think we have killed it. Then he realises with terror that he bears it within him:

> Not like a satisfied cow ruminating on its feet,
> But like the virgin mare, its mouth still burning from the salt
>     it has taken from its master's hand,
> How can he keep back and restrain that huge and terrible thing
>     that rears and cries out in the narrow stall of its personal will,
> When the smell of the grass comes in through the cracks in the
>     door with the wind at dawn?[12]

*

Man is wounded—a sign of his greatness, often hidden and always indelible. When the wound breaks the surface of consciousness, it assumes the most varied forms. It becomes

12. Paul Claudel, *La Ville*, second version, Act III (1911 ed.), p. 291.

the source of a continual unrest, of a deep dissatisfaction which not only prevents the sufferer from being content with any one position, but from being satisfied with progress in any single direction. It is the motive-spring of thought which drives him to break through the successive circles in which the life of the human animal tends to unfold, and disposes in turn of the critical systems and of the positivist wisdoms which seemed able to dispose of it. It may take the form of dread without any precise object:

*Aliis oppressa malis in pectore cura* (With heart weighed down by other evils).

This anguish cannot be described in all its forms and psychological expressions—not even 'depth'-psychology can reach it except in its manifestations. Sometimes it is a presentiment, the presentiment of another existence, and those who experience it vividly can sometimes communicate its flavour, or at least the suspicion of it, to those around them, thanks to the connivance of the spirit which is found everywhere, though it is almost always dormant and subject to the mysterious laws of germination. It has been called 'the appeal of transcendence'. . . .

One might try to give comforting explanations of this universal phenomenon. Equally it is possible to criticise many of its cruder manifestations. One may condemn its many distortions and point out many counterfeits of it—which are all the more serious when they hinder the normal growth of the spiritual life. One could go on almost indefinitely unravelling the confusions which the undeveloped mind entertains on this subject.[13] It may be further observed that many people who

13. M. Gabriel Marcel is right to observe that if 'the need for transcendence presents itself above all . . . as a kind of dissatisfaction . . . the converse does not seem to be true, it does not seem that one would be in the right in saying that every kind of dissatisfaction implies an aspiration towards trans-

are on the whole satisfied with themselves have not the faintest
idea of it; whereas it is often remarkably clear in certain
states of illness or when the social organism is not in sound
health. But it would be a very poor observer who thought
that it was just an anomaly, a passing disease, a sort of excre-
scence which could be removed altogether one day, a phantom
of the mind which could be dissipated, a strange voice which
could be reduced to silence. It would be most unrealistic to
imagine that physical or social health or the progress of science
were the cure. That would be to misconceive all that is most
human in the human being and which 'makes him more than man'.

Let us suppose, however, that the cure has taken place. We
should have no hesitation in preferring worse health if such
good health were to condemn us to a complacent humanism,
if the balance reached left man glutted, and he no longer
regarded himself as a problem. What a depressing ideal it
would be—a terrestrial existence undisturbed by struggle or
contradiction, without suffering but also without aspiration
and untouched by the search for the Absolute! A perfectly
ordered world with no room for either saints or heroes! An
ideal world, perfect in its circumscribed reality, completely
adapted to its surroundings, where there is an exact balance
between the objective and the subjective, where man's idea
of himself and his concrete existence are identical,[14] so that

cendence' (*The Mystery of Being*, I, English trans., p. 42). This should put us
on our guard against a romantic illusion, and equally against the opposite,
anti-romantic illusion.

14. Cf. Henri Niel, *Athéisme et Marxisme* in *Lumière et Vie*, 1955, p. 78: In
Marx's view, 'religion is born of the dualism between the idea and reality,
between the idea which man forms of himself and his concrete existence.
As long as that hiatus exists there will be religion. Marx's aim was to
put an end to that sense of insufficiency and lack of adaptation by realising
a world where man would feel really at home.' Cf. Emmanuel Mounier,
*Carnets de route*, ii, p. 415.

there would not be the smallest fault or crack through which to communicate the mystery of being, no further adjustment in the wonderfully balanced machine of the human universe, no room for man's struggles with himself or for a genuinely personal decision! One might go on using the words 'humanism', 'culture', and 'spiritual life', but in what a degenerate sense! And from the Christian point of view, what a monstrosity! Even from the merely human point of view, what misery it would be! Does the vast effort which carries us forward today lead to that prison-cell?[15]

But in fact we are not faced with such a dilemma. On the contrary, the truth is that certain social conditions, where the injustice or the misery is too great—although they favour, perhaps, certain crude aberrations—shut man off from the life of the spirit. So we should work with all our hearts and with no misgivings, and certainly without the slightest danger of going too far, to improve the lot of man and to promote progress on every front: success, however great, can never heal the wound. Even if 'the leap into the reign of liberty', which Engels foretold, were to be realised on earth, the wound would still remain. Our consciousness of it would only be keener and purer. What social disorder has not created, social order is powerless to cure.

\*

The belief that man's hour had sounded became general for the first time in our age; it came in like a racing tide: man was sufficient unto himself in his immanence and in his finitude, usurping the prerogatives of God. This was the madness

15. It almost seems as though the Marxists themselves were somewhat uncomfortable, and this perhaps accounts for the fact that they much prefer to draw an ideal portrait of the Marxist man in society as it is at present than to depict man in Marxist society.

of Kirilov, of Zarathustra and of Feuerbach, of the 'humanist' and the 'superman'. . . . It has rightly been called a 'tragic mistake'. Man excels, it is true, in transmuting the actual conditions of his misery, whether physiological or social, into all manner of dreams. There is certainly much truth in the opposed psychologies of Marx and Freud, to mention two great parallel examples. There is truth, too, in Comte's idea of a first 'theological' age and in analogous ideas in the worlds of our philosophers and historians. One of the signs of a mature spirit is without doubt to renounce false forms of transcendence, and all the luxuriant vegetation which draws off the sap and produces no fruit. But let us not forget the wisdom of the first great 'reductionist'. When old Xenophanes of Colophon uttered his seemingly sceptical words: 'If oxen and horses had hands and could paint and draw . . .', he was bent upon purifying, not destroying, the idea of divinity. Let us not forget either that the reality of nature and the reality of man, once recaptured, if they needed to be, still have to be explained, and also to be explored and penetrated, preserved and saved. We must be careful to see that when, like Xenophanes, we begin by 'reducing', we do not end by mutilating, and that the conquests of science, wrongly interpreted, do not confuse and cloud the mind, and that in ridding ourselves of one illusion we do not fall into another, its antithesis. For there is indeed an illusion of the absolute, but there is also an illusion of the relative; there is the illusion of the eternal, but there is also the illusion of the historical; there is the illusion of transcendence, and also the illusion of immanence; a mystical illusion and a positivist illusion. That is to say, if one misconceives the relative and the historical, one can, of course, only obtain a pseudo-absolute, a pseudo-eternal, and one's liberation is a dream. But, on the other hand, it is no less true that if we misconceive the eternal and the absolute, we are left with only a pseudo-historical, a pseudo-temporal, a path that does not lead to liberation. In short, 'mystification' takes place in more than one direction.

\*

There is a mystical or celestial illusion—and there is a positivist or terrestrial illusion. Let us call the one spiritualist and the other materialist. Now, they are not only individual illusions or illnesses. Either can stamp long periods of history. Normally the celestial illusion comes before the terrestrial, which is why the second is a double illusion, taking itself for critical sagacity. Yet it serves no purpose to dissipate one illusion if we fall into the other.

The man whose guide is the Gospel will be on his guard against both. The idea of transcendence implies immanence. The dogma of the resurrection and the biblical behest to work the soil are reliable guides; so is the precept of brotherly charity. The intellectual maturity and the technical progress of the last few centuries help us to deepen our understanding of it. We believe with St. Paul that '*the figure* of this world passes', and we refuse to sacrifice either side of this truth, recognising their solidarity. We do not want a spiritual life in a dream-world, nor an eternity which is not prepared for us by time. But neither do we want a closed humanism, an 'inhuman humanism'. 'Nothing but the earth' is the cruellest of all illusions.

\*

The heaven of the mystical illusion does not exist. But the earth of the positivist illusion, the temporalist illusion, does exist—and takes its revenge.

\*

We do not protest enough against the way in which the idea of God is distorted among Christians. We are always anxious to spare the weak; we avoid startling the weak-minded and keeping the impure away, in the hope that contact with the Church holds out a possibility of enlightening and converting them; and we forget that there are others, no less weak, the unbelievers, who are scandalised by our accommodations.

To allow the truth to be obscured is invariably a cause of scandal somewhere, even if one has been tempted to prevaricate in order to avoid scandalising this or that individual.[16]

*

No proof gave me my God, and no critique can take him from me. However acute it may be, that critique will provoke another. But perhaps it must first fulfil its salutary role. Without forcing me to make any concessions, it stimulates me to make progress. Without depriving my proof of value, it compels me to unearth its secret spring, to deepen and purify my faith.

Moreover, unbeknown to and in spite of himself, the atheist is often the greatest help the believer can have. Like the biblical Ecclesiastes, his criticism marks one of the stages of the dialectical process. He co-operates, unintentionally, in the 'purification of faith', which consists in 'freeing it more and more from the senses and human reasonings'. He provides the salt that will prevent my idea of God from petrifying and so becoming false.

*

The idea of God cannot be uprooted because it is, in essence, the Presence of God in man. One cannot rid oneself of that Presence. Nor is the atheist a man who has succeeded in doing

16. Jacques Leclerq, *op. cit.*, pp. 23–24.

so. He is only an idolater who, as Origen said, 'refers his indestructible notion of God to anything rather than to God himself'.[17]

*

Industrial civilisations are naturally atheistic, and agricultural civilisations are naturally pagan. Faith in the true God is always a victory.

*

As they gradually become more profane, modern civilisations expose us to the risk of losing God. Perhaps they will enable us to re-discover him at a deeper level, and the rediscovery may well prepare the way for new syntheses, without involving a return to earlier and indistinct ideas.[18]

*

No man without values—and no values which establish the value of man absolutely unless there is an Absolute which establishes them.

Man is of absolute value, because he is illuminated by a ray of light from the face of God; because, although he develops as he acts in history, he breathes the air of eternity. Unless that is true, any philosophy of man must be a mere vulgarity, a cynicism or an empty dream.

*

People imagined that by reducing everything to immanence, everything, beginning with himself, would be given back to man; on the contrary, it meant robbing him of everything he possessed and 'alienating' him absolutely. For it implied reducing everything to *duration*—to a duration with-

17. Origen, *Contra Celsum*, ii, n. 40 (PG, xi, 861B).
18. Cf. van der Leeuw, *L'homme primitif et la religion*, pp. 207, 167 f., 187.

out an eternal foundation, so that despite all one's efforts the moments of which it is formed disintegrate into fragments or add up into a mass but never form an interlocking structure. It was even thought that 'the treasures wasted in the skies' had been saved; that the Absolute belonging to a dream-world had been brought back to earth; but it was not brought back from God into Man; it collapsed into the relative, carrying the whole of man with it.

*

The world is the real work of a beneficent God and has a real value  It is not just the stage on which man has to act and choose, nor is it simply an instrument for him to use; it is, so to speak, the stuff of the world to come, the matter of our eternity. Man's task, therefore, is not so much to liberate himself from time as to liberate himself through time. His task is not to escape from the world, but to raise it up. Only, in order to understand time and the world, it is necessary to look beyond it: for it is its relation to eternity which gives the world its consistency and makes time a real becoming. And it is the hope of radical and final transformation which saves our terrestrial effort from futility.

*

Humanism, it has been said, 'is a fully articulated anthropocentrism which starts from the knowledge of man and purposes to give him value, thus excluding everything which alienates him from himself, whether it does so by subjecting him to supra-human truths or powers, or by disfiguring him by putting him to some sub-human use'.[19] But if the refusal, as it is called, to subject man 'to supra-human truths and powers' is a refusal of God and of divine truth, it would very

19. C. Brunold and J. Jacob, *De Montaigne à Louis de Broglie, introduction à l'étude de la pensée française contemporaine* (1952), p. 4.

soon lead to the disfigurement of man 'by putting him to some infra-human use'. The guarantee of man's value is to be found above his own sphere. The way in which humanism, that regards man as the supreme value, 'gives value to man' ends by resembling the exploitation of land or live-stock.

\*

Man has a twofold character; there is the historical aspect and the inward aspect, and the one cannot be dissociated from the other. He possesses the one by virtue of the other; but for the fact that his historical character is real, fruitful and clearly orientated, his inwardness would only be phantasmagoria, a psychological epiphenomenon; were it not for his substantial inwardness, his historicity would disintegrate in time, itself reduced to dust. . . . Man makes himself in and through history—that can be affirmed without presupposing any particular theory of 'progress'—and that is why each generation only fully understands itself as a link in the chain of humanity on the march. But the march of man would have no meaning, or rather humanity would not be on the march, and the very name by which we designate it would be a mere *flatus vocis*, if there were not, at the very heart of the world, an Eternal, drawing us to an End, impressing upon each one of us the seal of his image, and conferring upon each of us his unalterable inwardness.

\*

If man wants to find himself, he must aim above and beyond himself. It is not enough for each individual to take on a task that transcends him; the same must be true of each generation and for each community, and in fact for humanity as a whole. Otherwise our successes will only be external and precarious, threatened each time by a more radical crisis of nihilism, and turning, in the end, against man himself. No human fortune is worthy of absorbing man's attention. Humanity can only find equilibrium and peace—an active peace, an equilibrium in movement—by keeping its gaze fixed above the earthly horizon, in being faithful to its divine vocation.

Man needs a beyond which can never be grasped, a beyond that always remains beyond. He cannot find himself without losing himself.

At each stage, the final solution of the human problem lies in adoration. It can only be found in ecstasy.[20]

\*

'Man surpasses man.' Many of those who regard themselves as 'humanists' in the most exclusive and strict sense of the word, are prepared to recognise the fact. No man, according to them, is worthy of the name without a *sursum*. The doctrines that incarcerate him in a nature already fully realised seem to them untrue and contemptible. They require a movement towards transcendence at the very heart of immanence. . . . But in what sense can that movement be efficacious? Dreams of a collective future are vain unless there is an eschatological Beyond, already present and active in the womb of becoming. Equally our dreams of spiritual development are vain if there is no Beyond which is both transcendent and immanent at the same time.

Those who do not recognise the attraction of a Transcendent Being in themselves have nothing to offer but a transcendent movement towards nothing.

\*

At one moment we are told that we make our heaven in the image of this—all too real—world; and at another that our

20. Cf. my *The Drama of Atheist Humanism* (London, 1949), Part ii.

desire creates, by contrast, a mystical region where all the signs are inverted, in order to escape the bondage of society here below and fly to the freedom of our dream-world.

But we know quite well that our God is different. We know quite well that he is the Living God. Our faith in him does not depend upon those processes, and the hope that he pours into our heart does not deceive us.

Indeed, we see the processes which are pointed out to us quite as well as others. The mind cannot do without analogies in its attempt to represent God—so the door is wide open to risks. The oppressed naturally tend to find in the sky a refuge from their hardships and the malign influence of man—and that involves risks of another kind. It is only too easy, too 'natural', to mobilise the divine in the service of social realism, or to call it to the rescue of a utopian subjectivism. The one can call God the principle of his revolt, the other the principle of his tyranny. But the believer is on his guard against both abuses. He does not allow himself to be the dupe of analogies or of contrasts. He does not deify the earth, or hypostatise a heaven without reference to it. And yet he knows full well that, in the end, only his hope in heaven gives relish to the earthly work that prepares him for it.

*

Man, they say, is alienated by his God. But alas, the truth nowadays is rather that he is alienated *from* his God, deprived of his ultimate wealth, the most precious of all, where he could re-discover the principle not only of his *having* but of his *being*.

How sharply the traditional formula has been thrown into relief in modern times: he who is in me is *more than I myself*! —a formula adopted confidently by many of those who live its truth without reflecting upon it, and which a terrible experience of emptiness has demonstrated anew!

Man without God is dehumanised.

*

There have been tyrannical gods—and there is the God who makes us free.

Tyrant gods, nowadays, do not, as a rule, assume the names of gods. They prefer pseudonyms. But their tyranny remains the same.

You reject faith in God as an intolerable 'theocracy'? Yet with every day that goes by it is surely increasingly obvious that this could only favour a 'mythocracy' more terrible still. Empty the heavens and they are at once occupied by an army of myths more compelling than hunger, more despotic than the worst despot. . . .

*Non habebis deos alienos coram me* (Thou shalt not have strange gods before me). Such is the everlasting 'precept of liberty'.[21] Those strange gods, those false and mythical gods, are the ones who alienate us—monsters that devour us like the human passions of which they are the hypostasis.

You made the true God resemble them and believed that in rejecting him you rejected them all with a single gesture. But that proud gesture was the result of a misunderstanding, and you failed to perceive that one must choose between him and them. The dark gods who vanish as the Sun of Justice dawns, and whom he keeps at a distance, return at once under other names.

Whether the names be new or old, the names of gods or pseudonyms, they invariably possess some characteristic of the man who adores himself in them, and makes himself his own slave.

21. Origen.

How far are we destined to go in that slavery before humanity, with one voice, begins to cry out, 'I hold out my arms towards my liberator!'

\*

God is rejected as limiting man—and people forget that it is man's relation to God that confers upon him 'a sort of infinity'. God is rejected as enslaving man—and people forget that it is man's relation to God that frees him from all servitudes, in particular the historical and social. God is rejected because he obliges man to ratify everything—and it is forgotten that it is this same relation to God that confers upon him 'an infinite capacity for rejection'.[22] Men reject God on the ground that he alienates man by his transcendence—forgetting that 'it is in the affirmation of transcendence that man finds his most authentic truth.'[23]

\*

A purely humanistic consciousness, which will not recognise anything outside man, can never be quite certain whether the impulse which drives it towards life is hope or despair. To say that man is everything is surely to say that he finds himself faced by nothing. . . . Can man be saved if we can have recourse to nothing beyond man?[24]

\*

Unless God is at the source a revolt and its ally, it invariably ends in servitude.

Whenever we say No, we imply that on a deeper level there

22. The expression is André Breton's in *Position politique du surréalisme*, p. 11.
23. F. Alquié, *Philosophie du surréalisme*, p. 211. Cf. Albert Cartier, *Le problème de Dieu dans la philosophie de Blondel* in *Giornale di Metafisica*, 1955, pp. 833–848.
24. F. Alquié, *op. cit.*, pp. 210, 211.

is a Yes which provokes and originates it; rebellion always implies an acquiescence which is both deeper and more free.

✳

One can never, even for one generation, bracket off, as in a parenthesis, the immediate problems of existence or the whole problem of destiny.

Humanity is always a present question, with its elementary needs and its passion for the absolute.

✳

The believer can hardly help being saddened at the sight of humanity caught in the quicksands—and who knows for how many centuries it may be?—at the very moment when it seems to aspire more fervently than ever to be free. He sees it shun its God as 'a strange being'.[25] He sees it alienate itself in the very act in which it believes it has finally freed itself. How can one avoid being saddened at the thought that it is lowering itself in a movement that seemed as though it were struggling to greater dignity?

In time, perhaps, those in whom the original aspiration lives on will see the danger. And then, perhaps, they will recognise that the believers they first took for adversaries are really their indispensable allies.

✳

*Homo sapiens* has again become *homo faber*—but this time he is the builder of a whole world, and so more than ever the builder of himself. He is no longer the hard-pressed animal, but a creator. All that is true enough, but does it not mean that

25. Cf. Karl Marx, *Das Kapital*: 'It has become practically impossible to ask if a foreign being exists, a being placed above nature and man, since the question implies the non-essentiality of nature and of man.'

once again he must go farther and rediscover a new wisdom?
And how can that be achieved except by a higher and richer
contemplation?

＊

In order to avoid acknowledging that I have received from
the Creator the traits which make me a man, shall I consent to
alienate them in favour of some future, or rather mythical,
Entity, which means nothing to me and cares nothing for me?
On the one hand, if I acknowledge the gift the inalienable
nobility of human nature is ensured, and if I sacrifice myself
for my brethren my sacrifice has meaning. On the other hand,
my consciousness itself is sacrificed, and as a result of a complete and final alienation, I am simply a cog in the vast machine
for producing, in the distant future, which I cannot know,
what is called, I know not why, Humanity.

＊

The idea of God within us is perpetually menaced with
extinction, but is always reborn. Everything threatens it with
ruin, for everything is a scandal to us, when lo and behold!
the very threat that menaced it with death gives it fresh life.
Each day brings a new witness of it. For man will never finish
wrestling with God. The mysterious struggle between Jacob
and the Angel, so foolhardy and yet so necessary, so necessary
yet so unequal, lasts through the night—throughout the night
of our sombre history.

'It is originally God himself', Bossuet says, 'who is brought
low and rises again for the human race.'[26]

＊

26. Bossuet.

*Sub nocte Jacob caerula*
*Luctator audax angeli,*
*Eo usque dum lux surgeret,*
*Sudavit impar praelium.*[27]

(Jacob in the dark night,
The bold wrestler with the angel,
Sweated in the unequal struggle
Until the breaking of dawn.)

✳

*Exaudi me, Domine, Deus meus, Illumina oculos meos, ne unquam*
*obdormiam in nocte.*[28]

✳

27. Prudentius, *Book of Hours*, Hymn ii.
28. *Psalm* 12 (13):4.

. . . Abraham, desiring to know what should come to him
through the blessing of his first father, inquired about the God
for whom he was to wait. And as, following the inclination and
tastes of his soul, he journeyed about the world, asking where
God might be, and as he grew faint, and ceased his inquiries, God
took pity upon him who sought him only in secret: he revealed
himself to Abraham by means of the Word, as though by a ray
of light, and made himself known. . . .

St. Irenaeus, *Demonstration*, xxiv

In giving us his Son, God gave us everything. By delivering
up to us his unique Word, he revealed everything to us. There is
nothing further to wait for after Jesus Christ.

St. John of the Cross, *The Ascent of Mount Carmel*.

No man hath seen God at any time: the only begotten Son who
is in the bosom of the Father, he hath declared him.

*Gospel according to St. John*, 1 : 18.

No man hath seen God at any time. If we love one another,
God abideth in us. God is charity: and he that abideth in charity
abideth in God, and God in him.

*First Epistle of St. John*, 4 : 12, 16.

God is greater than our heart.

*First Epistle of St. John*, 3 : 20.

Now this is life eternal: that they may know thee, the only
true God, and Jesus Christ whom thou hast sent.

*Gospel according to St. John*, 17 : 3.

That which was from the beginning, which we have heard, which
we have seen with our eyes, which we have looked upon, and our
hands have handled, of the word of life:—for the life was mani-
fested: and we have seen, and do bear witness, and declare unto
you the life eternal, which was with the Father, and hath appeared

to us—that which we have seen and have heard, we declare unto you, that you also may have fellowship with us and our fellowship may be with the Father and with his Son Jesus Christ. And these things we write to you, that you may rejoice, and your joy may be full.

And this is the declaration which we have heard from him, and declare unto you: That God is light and in him there is no darkness.

If we say that we have fellowship with him, and walk in darkness, we lie, and do not the truth. But if we walk in the light, as he also is in the light, we have fellowship one with another; and the blood of Jesus Christ his Son cleanseth us from all sin.

*First Epistle of St. John*, 1 : 1–7.

*Sed in hac quaestione Deum videndi, plus mihi videtur valere vivendi modus, quam loquendi.*

(But in this matter of seeing God, our manner of life seems to me more important than our manner of speech.)

William of Saint-Thierry, *Aenigma fidei* (PL, clxxx, 398c.)

VERE DIGNUM ET JUSTUM EST, AEQUUM ET SALUTARE, NOS TIBI SEMPER ET UBIQUE GRATIAS AGERE, DOMINE, SANCTE PATER, OMNIPOTENS AETERNE DEUS: QUIA PER INCARNATI VERBI MYSTERIUM NOVA MENTIS NOSTRAE OCULIS LUX TUAE CLARITATIS INFULSIT UT DUM VISIBILITER DEUM COGNOSCIMUS PER HUNC IN INVISIBILIUM AMOREM RAPIAMUR. . . .

# POSTSCRIPT

The attentive reader will have seen at once that there is nothing in this little book which has not been borrowed from the double treasure of the *philosophia perennis* and Christian experience. The author imagined that the same was true of the first two editions, which appeared under the title *De la connaissance de Dieu*. He was astonished, at first, to hear that this had been questioned by some readers. For such a doubt to be possible there must, however, have been some risk of misunderstanding. Though one might say of questions touching the knowledge of God what St. Augustine said of that knowledge itself: *Nomen quippe non sonaret aenigmatis, si esset facilitas visionis*[1] (The word would not sound enigmatic if we had the power of vision), or again what St. Leo said of supernatural mystery: *Inde oritur difficultas fandi, unde adest ratio non tacendi*[2] (The difficulty of expressing oneself arises from the same source as the need for not keeping silence). To speak of God is as dangerous as it is necessary.

The danger, however, is no excuse for silence. 'God's truth is at once so exalted, and of so delicate a nature, so to say, that human language cannot touch upon it without in some way wounding it. . . . Yet after all, if you wait to find words worthy of God you would never speak at all.'[3] Bossuet's wise words seemed to us to clinch the matter.

Nevertheless, the need to speak of God is not in itself an

1. St. Augustine, *De Trinitate*, XV, ix, n. 16 (PL, xlii, 1069).
2. St. Leo the Great, *Sermo 9 in Nativitate Domini*, i (PL, liv, 226B).
3. *Sixième Avertissement aux Protestants*, No. 38.

excuse for lack of skill. Whatever the 'intentions' of an author may be, not only in the ordinary sense of the word but in the purely intellectual sense of *intentio*, the *intentio* of his thought, the general meaning which his aim gives to his work, the direction in which it is engaged,[4] it is always possible that some abbreviation of thought, some elliptical expression or some word with more than one meaning, may put some reader on the wrong scent; the stress or the tone may be too weak at one point or too strong at another, and so may endanger the delicate balance of truth in some minds. It might be added that a discontinuous form makes greater demands upon reflection and makes a full understanding of the various formulae more difficult. That is why we have closely revised the text in response to well-intentioned and authoritative requests. Many precisions have been added, aiming at greater clarity. It was, however, impossible to ignore the fact that each time one touched on essentials, the supplementary explanation gave rise to new problems, so that the more one explained oneself the more explanation became necessary. The inevitable weaknesses of human nature will not be made a ground of complaint against us. Moreover, like the earlier editions, this edition, in its recast form, does not deal with all the problems treated in the classical works on 'natural theology'. Nor does it claim, any more than former editions, to be a substitute for them. At some points at least it has been expanded so as to complete, or to illuminate, certain passages which seemed to us important in themselves, or which seemed likely to embarrass certain minds. And at the risk of weighing down the book, some notes have been added with the aim of incorporating further explanations or necessary justifications.*

* As stated, some of these notes have been omitted and others have been curtailed in the English version.—*Translator.*

4. Cf. St. Thomas, *Quodl.* iii, a. 17, ad 1m.; *De substantiis separatis*, 12.

The character of the original work has, as a result, been somewhat modified. Our first intention had simply been to lend a helping hand to a few people in their search for God (and our reward has been that it was more than once accepted). Readers of this kind have little use, as a rule, for the citation of 'authorities'. Quotations, in fact, were reduced to a minimum, to a few specially chosen texts whose beauty or force seemed particularly telling. Now they are printed out or referred to, in footnotes.

It will henceforward be still more clear to all, we hope, that we attribute the same importance as the Catholic Church itself, as we said not long ago in similar terms, to 'the power of human reason, starting from created things, and without the help of supernatural revelation or grace, to demonstrate the existence of a personal God'. We do not confuse that power—presupposed by our whole endeavour—with particular concrete conditions in which it is put to use, and, for example, to recall the words already used, if the taste for God is one thing, we know that the proofs are another. There is not a page in this book which does not bear witness to our attachment—as profound, we dare say, as that of anyone—to 'the sane philosophy which we received as a legacy and as a heritage of long standing, from the Christian centuries' and, while recognising that it is right and suitable in a work which is not a manual of instructions, 'to disengage it from certain of its scholastic forms, less suited to the present time', we are very far from regarding it as 'an imposing monument, certainly, but belonging to another age'. It is the philosophy which nourished us, and our thought continues to live in that climate. We should like to be able to show that it is still richer and more nourishing, that it has more sap and is more fertile, than even its adepts imagine. Everyone will see, moreover, that here, as everywhere, we profess no indulgence for the sort of 'philosophical neurasthenia'[5] which seems to eat away the minds of a certain number of our contemporaries, and that we have no excessive leaning towards the 'novelties of the day' which preach an exclusive preoccupation with 'individual beings and the flux of life' or the simultaneous adoption of 'diverse doctrines'. In any case, we can safely leave to specialists the task of taking up the necessary discussions; our ambition has always been, and still is at this moment, simply to recall some eternal truths in a language that is not too antiquated. And finally, if we consider with all believers, that 'the teachings of the faith on a personal God and his precepts, are in perfect accord with the necessities of life', we do not regard this as in any way detrimental to their truth-value but, on the contrary, by virtue of it.[6]

I must now beg the reader's indulgence to draw attention to one or two special points.

One critic suggested that it was our design to 'return to the Fathers' in a sense which implied renouncing all the subsequent acquisitions of Christian thought. That was an error on his part; the present work should make this sufficiently clear. We attach great value to many of those acquisitions. The mania for novelties and for all forms of archaeological thought repel us equally, and we know full well how far they are from the spirit of Catholicism. If what is called 'a return to the sources' has sometimes given rise, in our time, to rash statements, it must be granted, in justice, that this is not our fault. On the other hand, it would surely be a novelty to regard the patristic contribution as simply obsolete, either in thought or in expression. Can it really be imagined that the

---

5. M. F. Sciacca: *L'existence de Dieu* (trans. Jolivet, 1951), p. 36.
6. All the passages in this paragraph between inverted commas are taken from the encyclical *Humani Generis*, Part III, 'The position of traditional philosophy in the Church.'

patristic tradition, which is still the source of 'spiritual life' in a narrow sense of the expression, is no longer of any use in our intellectual inquiries? Is it no longer fertile? Has everything it contained been completely assimilated, digested, systematised and 'surpassed' by subsequent speculation, and is it now a waste of time to turn to it? Not one of the great Christian thinkers down the ages would concur with that view. Their example is in the opposite sense. Thought does not progress like a technique. That sort of break, that sort of practical contempt, would surely be full of dangers. Were it necessary, the warnings so clearly set out in the encyclical *Humani generis* should suffice to preserve us from them.[7]

Another critic, in a much more moderate form, informed the author that 'even when he gives the impression of following St. Thomas faithfully, his thought develops outside the synthesis and the spirit of St. Thomas'. Perhaps that impression resulted from the fact that in the earlier text there were fewer references to St. Thomas than to subsequent systematisations. Let us translate the observation into more exact terms by saying that in fact our constant concern in this matter, as in others, was not to present St. Thomas as standing against the whole Tradition, but rather to throw into relief the traits in which that Tradition finds in him its most eminent witness. We do not regard the 'common Doctor' as an 'exclusive Doctor' who dispenses us from the task of familiarising our-

7. 'Both sources of divinely revealed doctrine contain rich stores of truth so great they will never be exhausted. That is why the sacred sciences are continually rejuvenated by the study of sources, whereas speculation which fails to promote the study of the revealed deposit becomes, as experience has taught us, sterile.' What is said here of theology in the strict sense of the word is no less true of the whole of Christian thought, and it would clearly be contrary to the spirit of the encyclical to exclude everything which concerns philosophy. Cf. Th. M. Zigliara, *Œuvres philosophiques*, vol. ii (Lyons, 1881), p. 12.

selves with the others; and we deem it regrettable that a certain partiality, inspired by a misguided strictness and artificial controversies, should sometimes have obscured the sense of profound unity which exists among the great masters—a unity which M. Gilson, himself a subtle analyst of their individual characteristics, recently recalled to mind.[8]

To regard this as eclecticism would be entirely false. In a work which is not in the technical sense a philosophical work but a series of free reflections on the most fundamental themes, such an attitude is not only legitimate: we regard it as necessary. It safeguards the unity of the *philosophia perennis*. Among other advantages, it allows the assimilation, as far as possible, of many thoughts whose significance or power of suggestion overflows the meaning given to them by their immediate context. Other studies, more scholarly or more historical, belong to a different category. But in addition to these, there are what we have called marginal notes, and these humbler efforts should at least be tolerated alongside others, since they may sometimes have more chance of answering the needs of a certain number of minds. It is well to attend to the serious objections which are so often brought against us: not in order to give in to them, but in order to answer them; not that we should be intimidated by them, but because we must face them honestly. Those who are groping their way should be treated with respect and with sympathy. It is a mistake to conclude hastily that the truth could lose by so doing; it can sometimes even gain. And one should make a real effort to remember that God does not belong to a few professionals.

In fact, it should be added that the traditional philosophy is not exactly what certain over-simplified *exposés* might lead one to expect. St. Thomas himself, 'the most intellectualist of

8. Gilson, *L'esprit de la philosophie médiévale*, 2nd ed. (1944), p. 356, note (English trans., p. 379 f.).

Christian philosophers',[9] offers 'a constant resistance to the threats of rationalism'.[10] He requires 'that one should institute a severe criticism of our knowledge concerning the things of God'.[11] His negative theology is not the anaemic and timid theology of so many modern 'spiritualists'. His doctrine of analogy, often wrongly understood, has more than one aspect: it is not the milk-and-water theory one finds here and there whose sole aim seems to be to reduce the chances of vertigo. His criticism of the concept, in so far as it concerns our knowledge of God, is far-reaching. The best interpreters have shown this very clearly: for example, M. Gilson in his fine book on Thomism, or, among others, Father Sertillanges who praises its 'audacity, which is as tranquil as it is liberating',[12] and who, commenting on one of the texts of the *Summa*, once allowed himself an exclamation of amazement: 'What', he asks, 'is this unbreakable unity, so rich and so full that our concepts approach it from all sides and are swallowed up in it.'[13] . . . It is easy to take comfort in the thought of the classical distinction between the meaning of a concept and the mode of its meaning, and in principle nothing is more just[14]; but it

9. *Id.*, *Ibid.*, p. 36. Cf. the same author's *Introduction à l'étude de saint Augustin*, 2nd ed. (1943), pp. 112–125.

10. M. D. Chenu, O.P., *Introduction à l'étude de saint Thomas d'Aquin* (1950), p. 139.

11. L. B. Geiger, O.P., *La participation dans la philosophie de saint Thomas d'Aquin*, 2nd ed. (1953), p. 262.

12. A. D. Sertillanges, O.P., *Les grandes thèses de la philosophie thomiste* (1928), p. 52.

13. A. D. Sertillanges commenting on *Prima*, q. xiv, a. 4. Cf. Louis Bouyer, *Le sens de la vie monastique* (p. 172): 'In reaching him we leave far behind all that the mind can conceive. Not only all our imaginations, but all our concepts vanish at his approach' (English trans., p. 110).

14. Cf. F. Taymans d'Eypernon, S.J., *L'encyclique 'Hamani Genesis' et la théologie* in *Nouvelle revue Théologique*, 1951, p. 7.

is sometimes applied too materially, as though one flattered
oneself it was possible to lay the former aside and retain the
latter intact, as though one could, at least at the high point of
one's thought, conceive that *modus altior*, starting from our
human qualities, the *modus altior* in all its purity, which is
found in God and in God only. That is simply to re-establish,
in a roundabout way, an element of univocity in our analogical
knowledge which is in fact denied by it. It is to forget that, in
reality, the analogy is not in the concept but in the judgment,
that it expresses similarity and dissimilarity at the same time,
indicating a 'relation' (*ordo*, *proportio*) which allows us to
affirm the former while taking account of the latter.[15] Or else,
fearing quite rightly to have to admit that our concepts are
only approximate, people refuse, quite wrongly, to admit
their inadequacy which, without robbing them of their truth,
inevitably affects them.[16] Such excessive timidity comes,

---

15. St. Thomas, *Prima*, q. xiii, a. 5.
16. That is because the epithets 'inadequate' and 'approximate' or even
'inexact' are taken to be equivalents—a regrettable confusion (when it is
not merely a matter of words). The encyclical *Humani generis* uses more
precise language. It rejects the claim that the mysteries 'cannot be expressed
in true terms, but only in approximate and changeable terms which indicate
the truth to a certain degree, but which also necessarily deform it'; a
claim which is not so much a daring idea as a vague and inconsistent one;
not so much a desire for accuracy as a confusion. God himself, in his answer
to Moses (*Exod.* 3:14), 'reminds us that all our statements about him are
inadequate' (A. M. Dubarle, O.P., *La signification du nom de Yahweh* in
*Revue des sciences philosophiques et théologiques*, 1951, p. 18). Maréchal
explains, as many others have done, that the 'signification' of the divine
attributes, that is to say the objective value which the affirmation in judgment confers upon them, rests upon a 'very inadequate representation,
inadequate because it is borrowed from our experience of creatures'
(cahier V, p. 234). Cf. A. D. Sertillanges, *St. Thomas Aquinas* (4th ed.,
vol. I, p. 404): 'The doctrine of analogy makes it possible to attribute to the
divine names a value which is positive, though inadequate.'

perhaps, from a lack of sufficient regard for the compensating elements which ensure the equilibrium of the traditional doctrine. To tell the truth, the reluctance which such attempts reveal derives from a pragmatic rather than an intellectual concern, and it is by no means certain that they give full due to the spirit of faith. Would everything be lost if one were not able to present God in tabloid form?

A firmer conviction is justified in being less tentative in its approach. It is not tempted to stop half-way to the truth, and so sacrifice its respect for mystery to a cowardly instinct for security. That is because, however far it extends the scope of 'negative theology', it knows full well that the solidity of the first affirmations which support that theology remain undisturbed. It is in no danger of confusing the *démarches* of negative theology with the withdrawals or hesitations of agnosticism. It knows, as we shall see, that the 'no' which follows on the 'yes' is not (to talk the Sartrian jargon for once) 'annihilation': the 'yes' lives on secretly within the 'no' as its necessary correlative; it orientates, determines and qualifies it. Even if everything suddenly seems to have been engulfed, it knows that nothing is lost. It knows, with St. Thomas, that the *remotio* is the fruit of the *excessus*. And it can say with St. Augustine: *Non parvae notitiae pars est, cum de profundo isto in illam summitatem respiramus, si antequam scire possumus quid sit Deus, possumus jam scire quid non sit*[17] (It is a part of no small knowledge, when we have emerged from this depth

---

17. St. Augustine, *De Trinitate*, VIII, ii, n. 3 (PL, xlii, 948). Cf. St. Thomas, *Prima*, q. lxxxiv, a. 7, ad 3m. 'If one must always end with a negation,' Father Xavier le Bachelet asks (*Dictionnaire de théologie catholique*, vol. iv, col. 1024), 'what would become of our knowledge?' But this agonised question is answered a little further in the same article (col. 1111): 'Not an absolute but a relative silence, and one which has its place not at the beginning, but at the end of our knowledge.' See Chapter V.

to breathe on that summit, if before we can know what God is we can already know what he is not).

Criticism, moreover, is not rejection. It would certainly be wrong to reduce intellectualism to 'a logicism which identified the ground of being with concepts. It would be a strange misunderstanding to confuse the idea, in its pure and luminous realisation, with the concept, that pale spark that the human intelligence extracts from the most obscure participations in the Idea.'[18] But it is not less true that the concept remains indispensable, and the truth it involves is not in any doubt; it only needs to be defined. The critique of the concept which we instituted, or rather which we recalled in certain pages, is also its justification—'for there is always more in the concept than the concept itself'—and we should be the last to wish to give in to the mirage of some other form of knowing as part of the normal life of the mind. It has been excellently said that the concept and the discursive method, by themselves, would undoubtedly build nothing but an unreal world; but there is 'a basis of intuition' in our knowledge which is implicit in them, and confers a real value upon them, while at the same time requiring them in order to express itself and perfect itself.[19] The 'natural knowledge' or the 'necessary affirmation' which we have discussed, awkwardly perhaps but certainly in the spirit of an ancient and unbroken tradition,[20] cannot be objectified otherwise than in concepts —although it always remains a living force at the mind's centre, and prevents it from settling down in the conceptual order.[21]

18. Charles Boyer, *L'idée de vérité dans la philosophie de saint Augustin* (1921), pp. 226–227.

19. J. Defever, *op. cit.*, pp. 107, 123.

20. It is contained in the well-known passage of St. John Damascene, *De fide orthodoxa*, I, i (PG, xciv, 789).

21. 'An intuitive movement', Defever writes (*op. cit.*, p. 125), which takes up and transcends the representation.

Interpreting it in this way, that is by refusing to admit that, in this world, natural man can have a direct 'intellectual vision' of Being, or intuition sufficient unto itself, and again rejecting 'innate ideas' in the proper sense of the word, even in regard to the first principles of the reason or the *prima intelligibilia*, we were in opposition to all the doctrines which tend to 'ontologism'.[22] Furthermore, we believe ourselves to have followed the essential scheme of Thomistic thought on this essential point, in preference to any other philosophy approved by the Church. And very certainly we were more faithful to St. Thomas than those who thought themselves in a position to criticise us on this matter. As Josef Pieper has recently reminded us, neo-scholasticism was no doubt not wrong in wishing to 'wash its master St. Thomas of the least trace of agnosticism'; but (provided one recognises at the same time the profoundly positive element of his inspiration) that should not involve ignoring the 'negative element' in his philosophy, most specially in regard to the problem of the knowledge of God. God known as the 'unknown': for St. Thomas that is the highest degree of our human knowledge.[23]

St. Thomas has 'described the imperfection of the instrument' which we must use in our search 'better than anyone else'. That imperfection, however, 'does not arrest its intrepid flight'[24]. That is because he, too, knows that there is something more fundamental in the human mind: not outside, but at the very heart of the intelligence. To banish from intellectuality the element which is neither form nor representa-

22. To whatever variety they belong, they are always characterised by the idea of a certain objective apperception of the Being of God.

23. *In Boetium de Trinitate*, i, a. 2, ad. 1m. J. Pieper, *De l'élément négatif dans la philosophie de saint Thomas* (Dieu Vivant, 20, 1951, p. 45).

24. J. Webert, *Saint Thomas d'Aquin* (1934), p. 48.

tion, that dynamic element, the movement of thought which is not the concept, since it explains the formation of the concept, but which gives it its soul—that would be to destroy 'intellectuality' itself, and to imprison the intelligence within the sphere of the relative. One may hesitate about its nature, or rather focus attention on one or other of its aspects, according to the problem involved, but—once one has perceived what an enigma knowledge is to itself, and the sort of questions which it sets—it is quite impossible to get rid of it. Everyone is free to desire a more 'clear-sighted intellectualism' as the rather misleading catch-phrase goes. But it is perhaps opportune to recall that it is not a matter of taste, and that the most clear-sighted intellectualism is not always the most authentic. Genuine intellectualism is not a narcissism of the concept. It is not the love of the intelligence for its own sake, or a complacent delight in its products: it is the free and confident use of the intelligence in search of the truth. It would be as well to guard against reintroducing, by some subtle deformation, a new subjectivism. Nor should presumptuous declarations be accepted too easily: it costs nothing to announce, with the aid of peremptory proofs, some definitive distinction or to present a thing as perfectly clear and without shadow of doubt; but, as has been very truly said, a philosophy is not judged by its promises but by its achievements.

An analogous preoccupation was at the root of yet another misunderstanding. Several readers failed to understand the prime object of the chapter in which the origin of the idea of God is considered. Enclosed within the sheltered circle of their scholastic disputes, they imagined in all good faith that these pages were written for them—against them, as they thought. Miraculously protected against the very sound of the assaults delivered upon our faith in God, they do not appear to have suspected for a moment the principal adversary which those pages had in mind. And yet that adversary is legion. It has proliferated for a whole century. Turn by turn it assumes the masks of ethnology, of sociology, of psychology and of the history of religion. It has invented a hundred different systems, from the animism of Tylor to the lucubrations of a certain school of psychoanalysis which denounces the grand illusion from which humanity needs to be liberated. It explains the whole idea of God in human consciousness by a series of transformations starting from dreams, belief in spirits, the mystification of language, cosmic fear, social alienation, etc., and confident that it has in this way established the 'genesis' of the idea—one might even say its empirical genealogy—concludes to its nonentity.[25] To contest that pretended genesis is not to profess belief in innate ideas, nor to undermine the value of ratiocination, of the rational operation by which we affirm God: on the contrary, it means giving that operation a free field. To extricate the affirmation of God from the meshes of an immanent 'dialectic', in which so many contemporary thinkers, Marxists and others, would like to enclose it and make it relative, does not mean, either, cutting it off from its logical foundations; quite the contrary, by removing it from the interplay of 'otherness and negation', the affirmation of God is given back its foundations and established once again upon the absolute. Was it not Engels who said: 'This dialectical philosophy dissolves any idea of absolute truth'? And is it necessary to remind philosophers that in present-day language the two words 'dialectic' and 'logic',

---

25. Thus Gustave Belot, after recalling certain 'modes of thinking' God and distinguishing various types, according to him irreducible in principle, of the idea of God, writes: 'The initial term of these complicated processes is posited by man's mythical imagination.' *Note sur la triple origine de l'idée de Dieu*, in *Revue de métaphysique et de morale*, vol. xvi, 1908 (2), p. 721.

far from being equivalents, are often opposed to one another? Only recently an intelligent critic and analyst of doctrines defined Thomism as 'a repudiation of dialectic'.[26] In any case, unless some attempt is made to resist the invasion of the dialectic when it goes beyond certain points, it becomes impossible to preserve the decisive purity of the *logos*. In the same way, when we showed how, in the course of history and of the evolution of religion, certain 'analogies' hardened and became fixed, we were very far indeed from bringing metaphysical analogy to book. We were simply reproducing one of the observations of the Book of Wisdom. We merely observed that, in certain minds or among certain peoples, sensible things such as the vault of heaven or the sun or lightning, through which divinity could be perceived as in a symbol, become at one point opaque, and instead of sustaining religion, imprison the religious impulse. Hence the diverse forms of 'naturism', which are only too easy to discover in history, and which some would have us believe to be the origin and final explanation of all religion.

The words 'analogy', 'dialectic' and 'genesis', in the context in which they occur, seemed to us to be clear enough in themselves for anyone who was to some extent in touch with the sort of problem which is met with everywhere nowadays. No doubt those who are misunderstood are always in the wrong. It would have shown greater wisdom to have been more emphatic in forestalling misunderstanding. And yet, if the misunderstanding was limited to a small number of people, and if it is, furthermore, quite easy to recognise what engendered it, perhaps the author of it may not be held entirely responsible. Perhaps those who fell into the misunderstanding might even be invited to examine their own attitude, to see whether it does not in fact make it quite impossible to fulfil

26. E. Borne in *Philosophies chrétiennes* (1955), p. 163.

a task which is unfortunately indispensable and thus prevent
the pressing recommendations of the Holy See from being
obeyed in any way.[27]

Reviewing this work with his usual sympathy, Father
Joseph Huby, since then entered into the Light of God,
expressed one regret. The knowledge of God through Jesus
Christ, only just mentioned at the end of the volume, is
nowhere examined. The lacuna is undeniable, and is, we
freely recognise, not without its disadvantages. By delaying
too long among the problems of natural theology, one does
indeed run the risk of forgetting how abstract the method is,
and allowing oneself to be caught in a sort of 'religious
philosophy' which usurps the place of religion itself. One
runs the risk of turning into an object of speculation, even if
contemplative speculation, the Being to whom one should
give one's faith—and give oneself in faith. This speculation,
it is true, cannot fail to develop sooner or later into negative
theology, for it is certainly true both that natural reason cannot enter into God whom it affirms and that the knowledge
of God *per negationem* is on any hypothesis the most perfect.[28]
But in a climate of unbelief, negative theology has a fatal tendency to drift towards agnosticism, if not towards an altogether
negative mysticism, or towards atheism pure and simple,
concealed only for a time. 'The definition of the Absolute',
Hegel said, 'can only be negative', and everyone knows the
end of the story in its living posterity. . . . Even supposing
the risk were less, it may seem at least that reflection obliges

27. Cf. the allocution of Pius XII *ad Patres Societatis Jesu in xxix Congregatione
Generali electores*, Castel Gandolfo, 17 September, 1946 (*Acta Apostolicae
Sedis*, 1946, p. 383).
28. Cf. among others, Father Maximilien Van Sandt (Sandeus), S.J.,
*Theologia mystica seu Contemplatio divina Religiosorum a calumniis vindicata*
(1627), pp. 89–125, particularly pp. 118–119.

us to discard any personal qualification from the Absolute as being imaginative or anthropomorphic, and then the mystery of the Divine tends to be substituted for the mystery of the living God, all the more hidden for being personal.[29] As all sense of the values which Christianity engendered in our consciousness is lost, people cease to understand that respect for the mystery of God becomes an avowal, all the stronger, of his Personality.

These dangers must not, however, be allowed to mislead us into overlooking the legitimacy, the necessity even, of a 'natural theology'. The history of ideas reminds us that, in fact, it needs the climate of faith in order to attain its proper balance. It was formed and developed by the great thinkers of the Christian tradition within the faith, though they may have treated it as relatively autonomous and stressed its rationality. That is manifestly true, as the latest histories of Christian philosophy have once again brought to light, and as the Council of the Vatican has fully explained.[30] That is the basis on which we, in our turn, have proceeded. No attempt has therefore been made, at least not directly, to fill the lacuna which Father Huby pointed out. That would have demanded a whole book and a direct appeal to faith. But the historical perspective which preponderates in the first chapter, and the concrete point of view which is ours on almost every page, do something, we believe, to remedy the defects which we have admitted. Moreover, how could the Glory which the disciples of Christ contemplated, fail to throw some ray of light here and there, if secretly, upon our path? How could we, or anyone, have abstracted entirely from all that the Christian revelation has

29. Cf. Ch. de Moré-Pontgibaud, *Sur l'analogie des noms divins. Au centre de l'analogie révélée*, in *Recherches de science religieuse*, vol. xlii 1954, pp. 322, 324, 328.
30. Constitutio *De fide catholica (Dei Filius)*.

definitely given us? When Jesus invited Nathanael to be reborn, he also invited the philosopher, in a sense, to the same meta-morphosis.[31] In principle the realm of reason and the realm of faith are quite distinct, and various affirmations can of course be classified accordingly without the smallest difficulty —as belonging to the one or the other. The mysteries of the faith remain inaccessible to rational investigation, while the authority and the laws of reason remain essentially unchanged in the believing intelligence.[32] Nevertheless, it is a fact that it is often a nice point to determine into which category great works matured by Christian thought should really be placed; it is open to question whether they are really philosophical or really theological. The discussion goes beyond not only the meticulous analysis of the text but also the general historical context of the work; for example, we have only to look at the literature on St. Anselm's *Proslogion* and on St. Thomas's *Contra Gentiles*. There is not, in fact, a single Christian whose philosophy would be in every respect what it is without his faith. And whatever some may say, that is eminently true of St. Thomas. The 'sublime truth', the keystone of his rational structure, is in the Bible, though one could not say that the Bible imposed it upon him, nor that his reason imposed it upon the Bible. His most rational thought derives part of its vitality from the soil of Revelation. It springs from the reli-gious life and flowers in a religious act. 'Dialectic and contem-plation are happily married in an exalted experience.'[33]

31. A. D. Sertillanges, O.P., *Le christianisme et les philosophies*, vol. i, p. 7.
32. Cf. Vatican Council, 3rd session, Constitutio *Dei Filius*.
33. Cf. E. Gilson, *Le Thomisme* (5th ed.), pp. 123–139 (English trans., pp. 84–95). M. D. Chenu, *op. cit.*, pp. 161, 275. See also A. M. Dubarle, in the article, quoted above, p. 20: 'One may ask whether this single name of Being would have contained the riches which Christian thinkers dis-covered in it if it had not been placed in the framework of biblical revelation,' etc. Cf. Chapter V.

We have certainly not sought in any way to imitate the manner or the tone of the great Doctor, nor of anybody else, any more than we have erased all trace of the 'conflict of thoughts'[34] which inevitably agitates the mind when it allows itself to be filled by the Mystery of God; nor have we attempted to exclude the personal coefficient from our reflections. Provided the substantial unity of doctrine is safeguarded and the adhesion of all to the teaching Magisterium is assured, there are still many mansions within the great Catholic family. There are diverse forms of exposition, answering to a wide diversity of temperament, itself willed by God. There are, furthermore, various historical situations with the needs which they imply. Fundamentally always the same, like the mind of man itself, the problem of the existence of God appears, in the course of ages, under new aspects which, even if it were not a necessity, would remain an obligation to take into account to the best of one's abilities if one wished to enlighten one's brethren. Whether, as some would have it, this is due to a deepening of the mind, or at least to an improvement of its technique, or whether, on the contrary, as others would say, it is due to a sickness of the mind, or whether, more simply, it is just a change in perspective, it is in any case a fact that the question-marks—the objections and the negations—do not occur at exactly the same point, nor with the same emphasis. Now it is they that dictate the starting-

34. 'Cogitationum conflictus': St. Anselm, *Proslogion*, prooemium. St. Thomas, *In Joannem*, c. i, lectio i, n. 1: 'intellectus jactatur hac atque illac. . . .' And who has not said with St. Augustine, *De catechizandis rudibus*, c. ii, n. 3 (PL, xl, 311): 'Mihi prope semper sermo meus displicet; melioris enim avidus sum, quo saepe fruor interius, antequam eum explicare verbis sonantibus caepero, quod ubi minus quam mihi notus est evaluero, contristor, linguam non posse cordi meo sufficere. . . .' Cf. *De doctrina christiana*, I, vi (PL, xxxiv, 21).

point.[35] Moreover, the believer's faith is not satisfied with a literal repetition in all cases. Without comparing this very modest and limiting essay with the giant efforts of our predecessors, we can safely say that it follows wholeheartedly in their train in the service of the same Truth.[36]

The Christian knows that the only way to a real encounter with God is the Living Way which is called Jesus Christ. It was that thought which suggested the French title of this work: *Sur les Chemins de Dieu*[37]—without implying directly, even about the first steps of natural knowledge, whether they are the ways by which we go to God or those by which God draws us to him.

35. St. Thomas, *In lib. I de Caelo*, 22, 2. *De perfectione vitae spiritualis*, c. xxvi, in fine.

36. Of the many authors to whom we are in debt, we particularly wish to recall Father Joseph Maréchal, S.J. (*d.* 1944), whose work has inspired many passages (though his thought has 'too often been simplified and perverted', L. B. Geiger, O.P., in *Revue des sciences philosophiques et théologiques*, 1954, p. 273), and, where the historical interpretation of Thomism is concerned, M. Etienne Gilson.

37. St. Augustine, *De civitate Dei*, XI, ii (PL, xli, 318). Cf. *Sermo* 117, *De verbis evangelii Joannis*, c. 10, n. 16 (PL, xxxviii, 670).